# Biotechnology

# Biotechnology

## Corporate Power versus the Public Interest

Steven P. McGiffen

LONDON • ANN ARBOR, MI

First published 2005 by Pluto Press
345 Archway Road, London N6 5AA
and 839 Greene Street, Ann Arbor, MI 48106

www.plutobooks.com



British Library Cataloguing in Publication Data
A catalogue record for this book is available from the British Library

ISBN 0 7453 1975 0 hardback
ISBN 0 7453 1974 2 paperback

Library of Congress Cataloging in Publication Data applied for

10 9 8 7 6 5 4 3 2 1

Designed and produced for Pluto Press by
Chase Publishing Services, Fortescue, Sidmouth, EX10 9QG, England
Typeset from disk by Stanford DTP Services, Northampton, England
Printed and bound in the European Union by
Antony Rowe Ltd, Chippenham and Eastbourne, England

# Contents

# Introduction

'If we don't play God, who will?'
James Watson[1]

Despite wide differences of attitude and approach in different parts of the world, it is possible to begin this study of the regulation of biotechnology internationally with some generalisations which I hope will prove useful in making sense of a complex picture.

Firstly, regulatory systems for the control of agricultural applications of biotechnology have been notable, in general, for their relative permissiveness when compared to the much more cautious approach which has prevailed in relation to medicine and health care. Agricultural biotechnology, through its effects on the environment, through changes in the molecular composition of familiar foodstuffs, and through a host of features which will be discussed in some detail in the next chapter, certainly has implications for human health. However, the fact that medical biotechnology deals directly with our bodies has focused the minds of legislators and made them less willing to give up any of their regulatory prerogatives to 'the market' or to any form of self-policing. Practical fears for the potentially disastrous consequences of mistakes or misguided initiatives, as well as unease at anything which smacks of the instrumentalisation of human beings, have dictated a relatively cautious approach.

Secondly, until recently regulatory systems for agricultural biotechnology have been based on an assumption that the way to ensure that it has no negative effects on human health or the environment is to assess each of its products separately, as if they were new only in the sense that, say, a new chemical compound with useful properties is new. The wholly novel nature of the process by which genetically modified organisms (GMOs) are created has been played down, ignored or even denied. The European Union (EU) and several other legislative authorities have recently departed from this tradition, creating a body of law aimed specifically at GMOs, their release into the environment and their placing on the market.

Thirdly, although ethical questions tend to be more to the fore in discussions of biotechnology in medicine and health care, agricultural biotech is also seen as impinging on vital moral concerns and therefore of raising issues which cannot be settled purely on the basis of practical considerations. This means that, for some participants in the debate, questions of safety, the enhancement of food production levels and quality, and other matters of immediate and obvious relevance to agriculture are not themselves definitive in determining whether one is 'for' or 'against' biotechnology as a means of developing agriculture. Ethical issues, bound up with political and economic questions, play a large role in deciding such matters.

Fourthly, each of these regulatory systems is being developed, though with varying degrees of willingness or reluctance, with a close eye on those being created in other countries. The mesh of obligations under which a modern state is placed by agreements presided over by the World Trade Organisation (WTO) are sufficient to ensure this, whilst in medical biotechnology in particular, international human rights law has also played a role. The WTO has, moreover, been added to and qualified by important agreements specific to the cultivation of and trade in GMOs, including the Biosafety Protocol and modifications to the United Nations general rules on food, the Codex Alimentarius.

Categorical statements concerning the safety of these novel techniques and their products are by definition the province of charlatans and their dupes. The precautionary principle, as it has been called, therefore provides the only rational basis on which legislative action can be based. The question is not whether there is sufficient certainty to proceed on a basis other than the precautionary principle, but whether the rewards offered by biotechnology are so great that we can afford to take risks to achieve them.

Concerns surrounding biotechnology go to the heart of discussions over control of the food supply, over relations between agribusiness corporations, farmers and consumers, between rich and poor and 'North' and 'South', forcing us to examine the whole future of commodity production. In the development of US agriculture since the Second World War, in the European Union's Common Agricultural Policy (CAP), and in the Green Revolution with its transformation of agriculture in much of the Third World, we can see certain common features: intensification of labour and land use; increased reliance on inputs from outside the cycle of production itself, as when manure or compost are replaced with chemical fertilisers, or 'artificial' pesticides are used to destroy pest species; and the consequent increased dependence of the farmer on the corporate supplier of such inputs, as well as of seed, specialised tools, and other necessities, a dependence enhanced by the preference for cash crops over those which can be consumed directly by the farmer and his or her family.

In medicine and health care, moreover, biotechnology raises similar questions, to do with the increasing commodification and expense of the business of staying well, and thus of control and availability. Problems of availability, in particular, and of resource allocation and management, are brought into sharp focus by what biotechnology has to offer to the sick and, as we shall see, to the 'not well enough'.

In addition, medical biotech has become the focus of intense ethical debate around a number of issues, from the status of the human embryo to whether an individual owns his or her own DNA.

It is with such questions that this book is concerned. It is neither a text book on international law, nor an academic work on economics, nor a deep analysis of ethics, though I hope it will be useful to students of law, economics, business studies and those subjects whose specialists deal with the kind of ethical and philosophical difficulties confronting men and women whose task it is to regulate biotechnology. Rather, it is an attempt to ask what problems the application of these have posed for regulators and what their solutions have been, as well as to understand the way in which biotechnology is transforming the world, and ask whether we wish to see the world transformed in this way and, if we answer no, whether there is anything at all which we can do about it.

Some legislators have responded intelligently to these questions, while others have preferred to ignore both the public interest and the science and bow instead before the furious, self-interested lobbying of multinational corporations with a stake in biotechnology. In five chapters, this book takes the reader on a tour through existing biotech-related law in the European Union, the United States, the rest of the economically developed world and, finally, those developing countries which have found themselves confronted with the issue. The European Union's new raft of legislation for the control of GMOs is looked at in some detail, while the following chapter deals with the United States and its in comparison extremely lax regime, a system based on the idea that the products of genetic modification are 'substantially equivalent' to organisms produced by traditional cross-breeding methods. In addition, the approach of the two great trading powers to medical applications of these new technologies is contrasted. In the chapter on other developed countries I discuss the way in which each is following the lead of either US or EU, but in every case with important differences of law and practice. Developing countries, on the other hand, are for the most part keeping their options open, with African countries resisting the attempt by the US to force them to accept GM food, while much of the Pacific Rim enthusiastically embraces medical biotech as a means to perpetuate a prosperity built on cutting-edge technologies. Finally, Chapter 5 is devoted to international agreements, including the breakthrough Cartagena Protocol which for the first time gives – albeit in the face of a lot of ifs and buts – national authorities the right to refuse imports of GMOs. Away from agriculture, the menace of eugenics and other inappropriate applications of biotech is considered in the light of existing and proposed instruments through which these might be prevented.

Biotech carries risks, and promises rewards. Simply put, do the latter justify the former? And, if we ask this question, and try, as citizens, to answer it not with a straight 'yes' or 'no' but in a measured and considered way which shows we are taking our responsibilities seriously – for, after all, it is supposed to be ourselves in this civic role who are the ultimate regulators – then will anyone heed our reply?

## FURTHER READING

This book is written by a non-scientist for other non-scientists. Where necessary I have tried to explain the science and technologies with which we are dealing. For those who wish to read more on the basics of biotechnology, the following are recommended:

Morton Jenkins *101 Key Ideas: Genetics* (London: Teach Yourself Books, 2000)

Steve Jones and Boris van Loon *Genetics for Beginners* (Cambridge, England: Icon Books, 1993)

Steve Jones *The Language of the Genes: Biology, history and the evolutionary future* (London: Flamingo, revised edition, 2000)

Colin Tudge *In Mendel's Footnotes: An introduction to the science and technologies of genes and genetics from the 19th century to the 22nd* (London: Vintage, 2000)

Eric S. Grace *Biotechnology Unzipped: Promises and realities* (Washington, DC: Joseph Henry Press, 1997)

Stephen Nottingham *Genescapes: The ecology of genetic engineering* (London: Zed Books, 2002)

Gina Kolata *Clone: The road to Dolly and the path ahead* (London: Penguin, 1997)

Martha C. Nussbaum and Cass R. Sunstein (eds) *Clones and Clones: Facts and fantasies about human cloning* (New York: W.W. Norton and Co., 1998)

Matt Ridley *Genome: The autobiography of a species in 23 chapters* (London: Fourth Estate, 1999)

# I
# The European Union

The construction by the European Union and its member states of legal frameworks governing the biotechnology industry occurred against a background of growing problems for the industry. Conflict with the United States over the labelling of food derived from GMOs, widespread unease over ethics, and scepticism among the electorates of most member states may have given the impression that the EU authorities have a generally negative view of the industry. Nothing, however, could be further from the truth. In both practical and ideological terms the EU is one of European biotech's biggest supporters, channelling public funds into its coffers, pumping out a stream of pro-biotech propaganda, and using its various powers to put pressure on member state governments to toe the line.

## THE BARCELONA DECLARATION

The European Union's approach to the biotechnology industry and its products was confirmed at Barcelona in March, 2002 when the European Council – made up of the heads of state and government of the member states – adopted the Commission's proposals contained in an official Communication, *Life Sciences and Biotechnology*, a virtual love letter to the industry officially placing biotechnology at the centre of Europe's economic future.[1]

'Europe is faced with a major policy choice,' wrote then Commission President Romano Prodi.

> Either we accept a passive role, and bear the implications of the development of these technologies elsewhere, or we develop pro-active policies to exploit them in a responsible manner. Life sciences and biotechnology are widely recognised to be, after information technology, the next wave of technological revolution in the knowledge-based economy, creating new opportunities for our societies and economies.

Claiming that by 2005 the European biotechnology market 'could be' worth over €100 billion and that by 2010, global markets for non-agricultural biotech 'could amount' to over €2,000 billion, the

Commission President bemoaned the way in which the EU was being outstripped in the sector by the United States. Over €2 billion of public money would be committed to putting this to rights. A regular 'Life Science and Biotechnology Report' would be produced, including a rolling work programme for related legislation. Legislation would be reviewed.[2]

## WHAT THE EU CAN AND CANNOT DO

The European Commission, however, has no right to enact legislation, but may only propose it for the consideration of the member states and European Parliament, and then only within certain carefully defined policy areas. These are laid down in the Treaty on European Union, which attributes 'competences' to the various EU institutions, leaving everything else to the member states.[3] As far as biotechnology is concerned, the extent of EU 'competence' depends on the industrial sector involved. For agricultural biotech, the role of the member state governments is generally limited to, firstly, their contribution to the decision-making procedure at the Council of Ministers, the body which directly represents them; and, secondly, their transposition, interpretation and implementation of Directives and Regulations.[4] In health care, however, the member states continue to run the show, with the Commission having only very limited competence.[5]

Thus, while it is possible to write about 'European' laws governing agricultural biotech, applications in the health care sector tend still to be regulated by national laws which vary hugely. There are exceptions: the 1998 Directive on the legal protection of biotechnological inventions, discussed below, covers all applications of biotechnology. In addition, embryo research would be covered by provisions for the use of and trade in human tissues and cells, contained in Directives adopted in 1998 and 2004.[6] More generally, the Charter on Fundamental Rights of the European Union, adopted in 2000, prohibits both 'eugenic practices' and 'the reproductive cloning of human beings', though its legal status remains a matter for debate.[7] The EU's research programme feeds directly into the health care and pharmaceutical sectors, while European regulation of the patent system is of increasing importance. Insofar as they constitute economic activities, moreover, biotechnological interventions in health care and medicine must conform to the general rules governing the single internal market.

This restricted competence means that the programme of actions based on the Commission's Communication is more limited than the scope of that document might suggest. Towards the end of 2003, the EU Council of Ministers adopted a 'roadmap' which limited itself to

> assessing future needs, disseminating best practices about initial teaching, continuing profession development and mobility, strengthening the relation between knowledge and the market, co-operation involving the Member States and the private sector in the case of research, the transposition and application of EU measures for intellectual property, promoting the transfer of technologies, co-operation between the Member States, the European Commission and financial institutions with a view to improving the financial framework for biotechnology

and 'guaranteeing society's participation (helping to make biotechnologies ethically acceptable)'. It also promised 'a better regulatory framework' but proposed no concrete measures.[8]

## THE STRANGE CASE OF EUROPE'S UNWANTED GMOs

When it comes to agricultural biotech, European Union regulators have a problem. Though the extent of the rejection varies, EU consumers in every member state have shown that they simply do not want genetically modified organisms in their food. According to a 2001 survey by the EU's own opinion pollster, the number of people who believe that 'food based on GMOs is dangerous' varied from 38 per cent in the Netherlands and 43 per cent in Finland to 89 per cent in Greece and 68 per cent in France, averaging out at 56 per cent across the whole of the Union. Moreover, adding the 'Don't Knows' to this figure gives us the answer to the question which *should* have been asked: an EU average of 83 per cent believe either that food based on GMOs is dangerous, or that it may be.[9]

The degree of opposition to GMOs, both active and passive, varies across the EU. In Britain, experimental trials were widely condemned, and not only by environmentalists, but by the British Medical Association and the Scottish Parliament. As one Scottish newspaper editorialised, 'the risks appear to have been played down and the questionable advantages played up as the corporations behind the technology, mostly American, play for high stakes and even higher profits'. Members of the Scottish Parliament were critical of inadequate risk-assessment and monitoring procedures and questioned the value of the trials.[10]

Tony Blair and his Science Minister, the unelected supermarket magnate Lord Sainsbury, on the other hand, acted as standard bearers for the industry, routinely rubbishing those who did not share their enthusiasm as 'Luddites' conducting 'a retreat into the culture of unreason'. Government-paid and corporate scientists were regularly trotted out, presumably in the spirit of encouraging the sort of debate the Prime Minister claimed to want, to back these splenetic attacks. One of these was Professor Philip Dale of the John Innes Centre. Dale is linked to Lord Sainsbury through the 'Gatsby Foundation', whose purpose is 'to exploit commercially scientific breakthroughs in plant science' produced by the Sainsbury Laboratory and the John Innes Centre, both in Norwich. Sainsbury had been head of the Gatsby Foundation before joining the government. Sainsbury's benevolence did not end there. Prior to becoming science minister he had donated £9 million to the Labour Party.[11]

### Growing resistance

Though Britain somehow combined the most popular, militant and determined protest with the most cravenly pro-biotech, pro-corporate government, the issue was generating increasing conflict throughout much of Europe. In Belgium, a vigorous movement, adopting much the same tactics as had that in Britain, emerged to challenge existing field trials and defend the EU moratorium, while a Green Party minister blocked new licences. France, too, in the wake of the colourful campaign led by farmer-activist Jose Bové and Greenpeace, and responding to doubts about field trials expressed in the conclusions of its own 'national debate', moved to tighten the rules, give elected local representatives a greater say in deciding whether and under what conditions trials should take place and enforce greater safeguards against contamination. Despite these sympathetic noises, however, GM trials went ahead in several parts of the country, where they were routinely trashed.[12]

In common with Britain, the Netherlands and France, Germany tried to sideline opposition by organising an entirely bogus 'national debate', timed so that it was virtually impossible for it to have any influence on the package of laws, agreed at EU level, which would actually govern the release of GMOs and marketing of GMO products within the Federal Republic. The debate had not even got under way when the German minister put her name to EU Directive 2001/18, which embodied many of the decisions which the national debate was supposedly held to determine.[13]

The Netherlands has seen little in the way of anti-GMO protest, though the left wing Socialist Party, which commands around 5 per cent of the national vote, campaigned outside supermarkets under the slogan 'Know what you're eating' and has produced leaflets explaining the issues. The Netherlands 'national debate' did produce a somewhat critical report from the Temporary Parliamentary Committee on Biotechnology and Food, but this was swiftly answered by the government, with an unequivocal defence of its own pro-GMO approach. In turn, the government was attacked by industry spokespeople who accused it of not doing enough and threatened to quit the country.[14]

Spain, the only member state where GM plants are grown commercially, also saw some opposition, with Basque farmers' groups, for example, protesting to the European Parliament that the 'use of GMO corn should be suspended...the moratoria, maintained on new approvals and...strict controls...introduced and applied should new GMOs be approved'. The farmers' group, the General Union of Basque Farmers, called, amongst other things, for an effective liability system.[15]

Opposition to GMOs was scarcely evident when they were first marketed in parts of the EU, including the UK, in 1994. By 1998, however, when it was announced that no further approvals for marketing or commercial growth of GMOs would be issued and the *de facto* moratorium began, widespread protest had emerged. In the meantime, most people in Europe had come to the conclusion that whereas GMOs might be dangerous, they offered few if any benefits and therefore were not worth the risk, however slight. The familiar scenario of 'shareholders get the profits, we get the risks' seemed to be playing itself out.

The fact that people had so recently and frequently been lied to – about BSE in the UK, contaminated blood in France, and asbestos everywhere – and that these scandals all emerged in a short period of time, created fertile ground for the anti-GM message of environmental activists. Organic food organisations and, particularly in France, groups representing small farmers, also began to organise against this new threat. These different but now converging lobbies latched on to public concern not merely over food safety and the environment, but about what might lie behind problems affecting these areas of universal concern. As one report on the formation of negative opinions of GMOs explains it, they 'are perceived as strengthening highly industrialized patterns of agriculture' which are seen in turn as being responsible for 'a variety of negative impacts' including 'pollution' and 'foot-and-mouth disease in the UK'.[16] A 1999 survey showed that only 3 per cent of Europeans trusted industry sources or political parties as a source of information on biotechnology, while the comparable figures for consumer groups (55 per cent) and environmental organisations (45 per cent) were much higher.[17]

It is not only a sceptical public which has expressed reservations about GMOs. Harder to dismiss as ill-informed are official scientific advisory bodies which have shared and reinforced this lack of enthusiasm. In July, 2003, a panel led by the government's chief scientific adviser David King produced a report which emphasised the need to assess every new GMO for risks to health and the environment, despite finding that none currently available posed any significant threat. In the same month, a report from the Environment Strategy Unit concluded that no available GM crop offered Britain any economic benefit.[18]

The real death blow for the image of genetic engineering came, however, with the final results of the field trials. Announced in October, 2003, the 'largest and most thorough of their kind in the world',[19] they showed that GM beet and rape, two of the three crop plants chosen, had a significant negative affect on wildlife, largely because they greatly reduced the number of weed seeds available. Maize was an exception, though this was challenged by environmentalists who argued that the design of the trials and interpretation of the data were flawed, largely because the wrong kind of weedkiller was used. GM oilseed rape (canola) showed a fivefold decrease in flora and a 25 per cent decrease in the butterfly population when compared to conventional varieties, while GM sugar beet significantly reduced the number of flowers growing in fields and on their margins.[20] These details aside, however, the massive trials showed conclusively that the industry had been wrong in its constant assertions that GM crops would not harm the environment. The massive assault on much of Britain's wildlife that came with the growing intensification of agriculture from the 1960s onwards, would be furthered by the introduction of genetically modified plants.[21] At the very least the trials showed, as even the pro-biotech *Financial Times* admitted, that the EU's moratorium on new cultivations had been justified, even if they continued to insist that 'a blanket moratorium is wrong' and to call instead for each crop to be assessed for its effects on biodiversity and the risk of cross-pollination.[22]

## FIRST STEPS IN REGULATING GMOs

The European Community passed its first regulatory measures specifically governing the growth and marketing of GMOs in 1990. The main legislation authorising experimental releases of GMO crops and the marketing of GMOs remained, until recently, Directive 90/220 which created an approval and licensing system. This was based on a step-by-step, case-by-case approval process which included a risk assessment designed to protect human and animal health and the environment. It was accompanied by a law governing GM micro-organisms, 90/219.[23]

Directive 90/220 gave member states the right to object to approval of a deliberate release by another member state. It was this provision which would eventually lead to a tailing off of approvals, culminating in 1998 in the *de facto* moratorium which would make the Directive's replacement necessary.

### Laws governing genetically modified micro-organisms

Directive 90/219 was superceded in 1998 by an updated Directive, 98/81. 98/81 governs the use of genetically modified micro-organisms (GMMs) in contained environments and their commercial use in the manufacture of various products, mainly pharmaceuticals and a few foodstuffs such as beer and cheese. It obliges member states to take 'all appropriate measures...to avoid adverse effects on human health and the environment which might arise from the contained use of GMMs'. GMMs are divided into classes, with different levels of safety precaution deemed appropriate in each case. Premises where GMMs are to be used are subject to an approval system. Before any 'contained use' begins 'where failure of the containment measures could lead to serious danger, whether immediate or delayed, to humans outside the premises and/or to the environment', a 'competent authority' designated by the member state must ensure that an emergency plan is drawn up, an exception being made in cases where such a plan has already been made at Community level. Some of the obligatory features of such a plan, including information to the public and to other member states, are outlined. As is customary, however, much more detail – on protection of workers, the public and the environment – is given in the annexes. Curiously, GMMs have been introduced into general use in food and other products with little reaction. Their use in contained environments appears to make them no more dangerous than other potentially hazardous substances, perhaps. For whatever reason, they have become ubiquitous in food production and other industrial processes with no significant resistance.[24]

### Regulation 258/97 of 27 January 1997 concerning novel foods and novel food ingredients[25]

Used in prepared foods, GMOs were until recently covered by sectoral measures rather than by 90/220 or any other 'horizontal' legislation. The most important of these was the Regulation on Novel Foods and Novel Food Ingredients. This Regulation, which in common with 90/220 has been replaced by or subsumed by the new legislative framework discussed below, set out rules for authorisation and labelling of novel foods including food products containing, consisting of or produced from GMOs. Under Regulation 258/97, however, there was available a so-called 'simplified procedure' which applied to foods derived from GMOs but no longer containing them, and which were therefore defined as being 'substantially equivalent' to existing foods. In effect, for such foods no authorisation was needed, though 'substantial equivalence' had to be established to the satisfaction of the Commission. As for labelling, the Regulation required all novel foods containing or consisting of live GMOs to be labelled, but non-living foodstuffs made from GMOs need not be labelled if they had been judged 'substantially equivalent' to an existing non-GM foodstuff. In addition, where GMOs used in the processing of food were no longer detectable in the final product, as would be the case, for example, with refined oils or sugars, these need carry no label. It was these features of the Regulation that led member states to the view that it no longer provided an adequate means to control the marketing of GMO-based foodstuffs.

The Novel Foods Regulation was supplemented by Regulation 1139/98 'concerning the compulsory labelling of certain foodstuffs produced from Genetically Modified Organisms'.[26] This was necessary for technical reasons because certain GMOs had already been on the market before the Novel Foods Regulation came into force and thus required special legislation to be enacted covering the soya and maize varieties which had been approved before the new Regulation came into force.

The debate around the Novel Foods Regulation raised many of the issues which would crystallise into the various hostile camps as the EU sought, at the beginning of the twenty-first century, to generate a wholly new framework of laws governing agricultural biotechnology and its products.[27]

## FROM FIELD TO FORK: THE NEW LEGISLATIVE FRAMEWORK FOR GMOs

In the early 1990s, industry mouthpieces criticised the EU's regulatory system for GMOs as 'a lead ball on our feet compared with elsewhere in the world',[28] as well as 'muddled, overly cautious, bureaucratic and expensive', arguing instead for a 'flexible, pragmatic' approach such as that in the US.[29] Judging by its 1994 discussion document aimed at improving the competitiveness of Europe's biotechnology industry, the European Commission agreed.

The Commission proposed the adoption of a 'simplified procedure' for the release of GMOs, arguing that restrictions should be relaxed because releases up to then had been shown to pose no problems. The relaxation would include a fast-track procedure for approving GMOs designed to express their own pesticides, removing them completely from the 90/220 system and administering them instead under the system for controlling chemical inputs in agriculture.[30]

Opponents of GMOs were up in arms over this attempt to weaken controls after only four years of experience. Experts who were not opposed to genetic engineering as such nevertheless warned that the move was at best premature. The head of the working group of scientists which monitored releases for the UK government's Advisory Committee on Release of GMOs into the Environment pointed out that even if 'up to now most modified plants have been similar to existing crops, and have not posed unusual risks' this should not be taken as a signal to weaken the law. Insofar as 'the whole point of making genetically modified plants is that eventually they will not be the same', releases were likely to become riskier. Any weakening of regulations might make it harder to stop research once it had got under way, even if unforeseen problems arose. Predictably the agbiotech industry itself reacted aggressively to defend these results of its intense lobbying, with a spokesman for SAGB, EuropaBio's predecessor, arguing that 'We need to move farther and faster.'[31]

### Moratorium

For once, however, the biotech industry's desire for a corporate free-for-all was to be thwarted. In fact, by the mid-1990s things were moving very much in the opposite direction, with consumer scepticism rising and legislators being pushed into taking their responsibilities seriously. When, in 1994 the Belgian company Plant Genetic Systems (PGS) applied for approval to sell herbicide-resistant oilseed rape in the UK, Denmark, Sweden, Italy, Austria and Germany refused to agree unless they received a commitment that the harvested seed would be labelled as genetically modified. Under the terms of the Directive, the Commission had the duty to propose a way out of this impasse, but 18 months after the application was received they had still not done so. Then, towards the end of 1995, Monsanto found itself in the same situation as PGS, with, however, one crucial complicating factor: the soya beans that Monsanto wished to market in the EU were grown in the US, and, because it had never been required to do so by law, the company had made, after harvesting, no attempt to keep GM and non-GM beans apart. Because they were selling them for eating rather than planting, they saw no reason to label them, arguing that there was no substantial difference between the two products. With the support of the US government, Monsanto claimed that this made it impossible for them to label the beans for export. When Denmark suggested that shipments carry a label to the effect that they 'may include' the GM bean, the US responded that this was 'not acceptable to US trade'. The spat was followed closely, and with an anxious eye, by the increasing number of US farmers growing crops for which GM products were now on offer.[32]

At the end of April, 1996, the beast that would become the EU's moratorium on new GMO releases stirred: for the first time since the adoption of 90/220, advisers to the Council of Ministers recommended rejection of an application to place a GMO on the market. The application, from Swiss-based multinational Ciba-Geigy, was for a variety of Bt maize aimed at the European corn borer, a major pest.[33] The firm had already received permission to market it in the US and Canada. The officials of the four member states which wanted it rejected had no problem with its ability to kill the corn borer, but they were perturbed by the plants' antibiotic 'marker' – introduced so that genetic engineers could test and monitor their own handiwork – and its linkage to a promoter gene[34] that made it more likely that it would be taken up by bacteria, making them immune to the antibiotic in question. This was, moreover, ampicillin, which was in widespread use by dairy farmers, while the promoter differed from previous promoters in not being specific to plants and therefore carrying a greater danger that it would be taken up by cattle, or some organism living in their gut. The fear was that the resistance gene might make cows immune to treatment or worse, through presence in their milk or meat, present a danger to human health.[35]

From the time that 90/220 came into force until 1998, the commercial release of a total of 18 GMOs was approved, though only two of these were for GMOs intended for food for human beings – a variety of soya and a variety of maize. Yet in October of that year, with 16 applications pending, the Council of Ministers announced a moratorium on new approvals for marketing or commercial cultivation.[36]

With both opponents and supporters of GMOs dissatisfied, it was clear that a new legal framework was needed. The Commission's strategy was to attempt to devise a system which would be acceptable to all parties, or almost all. Given strict regulation, those remaining unconvinced that GMOs can safely be released into the environment would be isolated from mainstream opinion. With new legislation, the EU was hoping to round up the usual suspects – environmentalists, radicals, jugglers and fire-eaters – and corral them away in the paddock of protest where they could be safely ignored and where they would not spook more docile members of the herd. The problem was that, as more information on GMOs became available, and to a background of food scares and scandals, the call for the technology to be scrapped was rapidly breaking through the fences and becoming mainstream. The industry in Europe, and would-be importers, faced a coalition of forces including radical environmental activists, groups such as Greenpeace, conservative Austrian farmers who had discovered that the GM-free status they could guarantee was opening up new markets, states with no immediate economic interests at stake, and a very large slice of a public increasingly sceptical of official and corporate reassurance.

It was against this background that the European Commission issued its *White Paper on Food Safety*. The White Paper proposed a major shake up, recommending the creation of a European Food Safety Authority to oversee food safety within the EU's internal market. As far as GMO-derived foods were concerned, a comprehensive programme of legislation was proposed. New measures would be based on an acceptance that GMOs were different, and needed special legislation to govern their growth, marketing and use. Balancing this was a clear commitment to the future of biotechnology in European agriculture. Get the law right, the Commission believed, and GMOs would be accepted.[37]

**Directive 2001/18/EC on the deliberate release into the environment of genetically modified organisms[38]**

The first element in the new legal framework would be the revision of 90/220. This measure was based on an explicit recognition that, contrary to official opinion in the US, GMOs are new products and should be treated differently to other foodstuffs. It established a 'step-by-step approval process on a case by case assessment of the risks to human health and the environment before any GMO or product consisting of or containing GMOs...can be released into the environment'. It does not apply to products 'derived from' GMOs – the Commission's notes on the measure give the examples of paste or ketchup from GM tomatoes – which continue to be covered by sectoral legislation, including the Novel Foods Regulation. Although 2001/18 does not exclude micro-organisms, where these are used in closed systems they continue to be governed by 98/81. GM animal feed is included, but rules governing the use of material derived from GMOs in feed would have to wait for the complementary measure on GM food and feed.

2001/18 tightened up the authorisation procedure, introduced a statutory obligation for member states to keep a register of GMO releases which must, along with other relevant information, be available to the public, and established rules for labelling and traceability of GMOs and GMO-derived foodstuffs. In addition, it required monitoring of long-term effects of the interaction between a particular GMO and the environment, and limited approvals to a period of ten years. When a GMO is approved by a member state the Commission must formally propose the authorisation, which can be adopted or rejected by the Council by Qualified Majority, meaning that a weighted minority of member states can block approval.[39]

The Directive requires member states 'in accordance with the precautionary principle, [to] ensure that all appropriate measures are taken to avoid adverse effects on human health and the environment which might arise from the deliberate release or the placing on the market of GMOs'. Each time a GMO is deliberately released an environmental assessment must be conducted. If the release is for purely experimental purposes, with no immediate intention to market the GMO being tested, a notification must be submitted to the member state where the release is to be made. The notification must give details of the GMO, the way in which it is likely to interact with the environment, under what conditions the release will be conducted, how it will be monitored for effects on the environment or human health, how emergencies will be dealt with, remediation methods and waste treatment, and the results of a prior environmental assessment. The possible results of any change in circumstances affecting the released GMO must also be assessed. If the GMO has been released under similar conditions elsewhere, there is a 'streamlined procedure' which allows a short-cut to some of this. Finally, the public must be informed, though commercially sensitive information may be withheld.

Anyone wishing to place a GMO on the market must first carry out field tests. This means that products intended for commercial rather than for experimental purposes must undergo both this procedure and then, should their field tests prove in their own estimation satisfactory, the procedure for placing on the market. Want to market a GMO in, say, Belgium? Having carried out your field tests (for which, of course, you must have followed the procedure for deliberate release, which itself requires the permission of the competent authority in the member state where the tests are scheduled to take place), you must then apply to Belgium's competent authority, a public institution, whether specifically created or considered suitable for the task, nominated by Belgium's government. The application must give a range of detailed information regarding the characteristics of the GMO in question: its likely interaction with the environment; 'pathogenicity: infectivity, toxigenicity, virulence, (and) allergenicity', amongst other potential problems; vectors used;[40] 'antibiotic resistance, and potential use of these antibiotics in humans and domestic organisms'; methods of genetic modification used; history of previous releases or uses; information on the site of the release; 'methods and procedures to avoid and/or minimise the spread of the GMOs beyond the site of release or the designated area for use'; 'methods and procedures to protect the site from intrusion by unauthorised individuals'; waste treatment; and emergency response plans.

If, once you have provided all of this information, Belgium approves, the rest of the EU's member states are informed and given the opportunity to comment, raise questions, or object. If no objections are made, consent is then given for the GMO or GMO product to be marketed throughout the Union. Where appropriate, conditions governing this marketing may be stipulated. If, however, objections to marketing approval are raised, the European Food Safety Authority (EFSA) is consulted. If, in EFSA's opinion, the objections are unjustified, or if they mean merely that certain conditions must be placed on the GMO or GMO product when it is marketed, the decision is handed to a regulatory committee consisting of representatives of the member states. If they decide that it should be given the go-ahead, the GMO or product can be marketed. If not, their decision can only be overturned by a unanimous vote of the member states at the Council of Ministers. If permission is granted, it is for a duration of ten years, after which it must be renewed. It will be monitored during those ten years and any problems taken into account if and when renewal is applied for. In every case any possible ethical aspects must be considered. At any time, if a member state has grounds for considering that a GMO or product constitutes a risk to human health or the environment, it may forbid or restrict its use or sale. Such a ban or restriction is, however, provisional. The European Commission and the member states must be immediately informed and a decision taken within 60 days as to whether it is legitimate.[41]

The Directive's recognition that a labelling system based on traceability through accompanying paperwork is needed was based on the fact that, firstly, reliable tests for determining GM content are not in every case available; indeed, where we are dealing with products which contain no trace of the GMOs used in their manufacture – such as refined sugars and oils, which contain neither DNA nor protein, and therefore nothing which has been modified – they never will be. According to the measure's opponents, such a system is inherently open to fraud. In addition, the numerous conditions 'make commercial release of any GM crop in the EU a lengthy, and therefore possibly unappealing, endeavour'. The European Commission's answer to this was to create a European Network of GMO Laboratories (ENGL) whose task is to ensure that GMOs and products derived from them can be traced through the food chain, backing up the documentation-based system with the use, where possible, of scientific methods of detection.[42]

Under 2001/18, its predecessor 90/220, and the Novel Foods Regulation, by mid-2003 a total of 16 GMOs could be legally marketed within the EU, while ten more awaited approval. The 16 included soya beans, maize, cottonseed oil, while those awaiting approval also include sugar beet.[43]

The moratorium held until 19 May 2004, when the European Commission approved, under the Novel Foods Regulation, the sweetcorn Bt-11 for human consumption in fresh or canned form. Almost immediately, however, its makers, Syngenta, announced that they would not commercialise the product, citing consumer resistance. Moreover, this breakthrough was achieved only because the Commission has the power under the Regulation to authorise a product without consulting the member states. Between December 2003 and September 2004, the Council of Ministers failed on eight separate occasions to vote to approve the marketing or release of a GMO or GM product, throwing the ball back into the Commission's court. Despite this, the unelected Commission effectively overruled the governments of the member states when it gave the go-ahead, in July of the same year, to Monsanto's GM maize variety NK603, this time under 2001/18 and only for use in animal feed.[44]

#### The Regulation on genetically modified food and feed and the Regulation on labelling and traceability of genetically modified organisms[45]

The Directive was only the beginning of an attempt to create a legal environment in which GM crops can be grown and GM foods marketed. Six member states – France, Italy, Austria, Denmark, Greece and Luxembourg – had made it clear that it would not in itself be enough to enable them to lift the moratorium. Where 2001/18 had sought to address inadequacies in the existing directive on deliberate release of genetically modified organisms into the environment, it had left the completion of a system which would perform the same task in relation to the marketing of GMOs and GMO products to be completed by two Regulations. Together with 2001/18, they form one of the strictest systems of regulation of genetically modified organisms in the world, though one which leaves a number of gaps in its provision for the protection of human health, consumer rights and the environment.

The Commission's solution to the problem of 'guaranteeing' the safety of food and animal feed containing GMOs or derived from GMOs was to establish a chain of documentation stretching from field to fork. The major features of the two Regulations introduced to achieve this can be listed as:

- traceability: it should, in principle, always be possible to identify whether a product is or contains or is derived from a GMO through following a chain of documentation back to the farm or other production facility where the GMO in question began life; this would be aided by the assignment of a 'unique code' to each GMO
- labelling: products containing or derived from GMOs should be clearly labelled as such

- because it was impossible to guarantee absolute purity, a threshold for GMO presence should be established below which, provided prescribed procedures for avoiding it had been followed, a food producer or packager would not be liable to prosecution
- the abandonment of the simplified procedure for putting on the market GM foods considered substantially equivalent to existing foods
- the introduction of a single risk assessment process under the supervision of a new European Food Safety Authority
- marketing approval would be limited to ten years
- transitional arrangements for foods already on the market under the system established by the Novel Foods Directive.

According to European Commissioner for the Environment Margot Wallstrom, the proposals would 'ensure a high level of environmental and health protection and pave the way for a proper labelling system'. For her colleague, David Byrne, whose responsibility was for public health, they would 'ensure that the regulatory framework in the EU is up to the high standard consumers expect'.[46]

Not everyone was convinced. BEUC, the umbrella group which brings together all of the EU's main consumer associations, welcomed the fact that 'the two proposals move[d] a long way towards giving consumers the possibility, previously denied them, to choose whether or not to eat food and food ingredients derived from GMOs', but criticised the section on adventitious contamination and called for a commitment to further research 'into any unintended effects on human health or the environment or unintended changes in the composition of food'.[47]

The European Parliament was deeply divided. Not surprisingly, the centre-right European People's Party (EPP)[48] adopted the industry line: principally, that the thresholds were impractical, and that labels must only be required where the presence of GMOs was 'analytically verifiable using validated methods' and thus the proposal 'to require process-based labelling of foods, where such labelling is not analytically verifiable, will be difficult to enforce and will undermine consumer confidence and trust in the labelling system'. At First Reading, however, marshalled by the progressive, astute and determined Austrian Socialist Rapporteur, Karin Scheele, a Coalition of the Willing was formed which included the Greens, the United Left, most Liberals, and her own group, with only the British

Labour Party members, bullied by Blairite pro-biotech fanaticism (or, in a minority of cases, sharing it), voting against the line. This succeeded in effecting a number of improvements, though only some of these would survive the EU's legislative obstacle course.[49] The most important of the Parliament's amendments were:

- an explicit reference to the precautionary principle
- an instruction to member states to 'encourage and contribute to the drawing up of guides to good segregation practice to be applied by food operators in order to avoid adventitious... contamination of food by genetically modified material'
- a threshold of 0.5 per cent, in place of the 1 per cent proposed by the Commission, below which the presence of GMOs in food not labelled as containing such would not constitute a violation of the law, provided it was 'adventitious or technically unavoidable'
- provision for member states to ban or otherwise restrict the sale of food or feed as an emergency measure
- improvements to the public's right of access to relevant information.[50]

Attempts to include processing aids and foods produced from animals fed with GMOs were defeated, as were amendments seeking to include food containing enzymes derived from GMOs.

It was clear in November, however, when the Council of Ministers adopted its own 'Common Position' on the two proposals, that the Parliament would not be able to carry much of this to term.[51] The deal between the 15 member states was so difficult to broker that Belgium's leading Francophone newspaper described it as having been delivered 'by forceps', and it was unclear to the very last minute whether it would be stillborn.[52]

As well as fixing the threshold for authorised GMOs at 0.9 per cent – higher than that for which the Parliament had voted, but slightly lower than the Commission's proposal – the Common Position differed from the Parliament's First Reading Report in a number of important respects:

- the Parliament's call for a complete ban on the presence of non-authorised GMOs in food was replaced by a 0.5 per cent threshold, though a ban would be instituted after 3 years, by which time it was judged that GMOs not currently authorised would have had time to go through the necessary procedure

- there was no reference to co-existence of GM, conventional and organic crops
- there were smaller differences over the details of the authorisation procedure and the public's right to information.[53]

The agbiotech industry was left seething by an agreement which, whatever its weaknesses, demonstrates that a well-organised citizens' lobby can still achieve a great deal, especially when it has science, coherence and sheer common sense on its side. We should not exaggerate this. GMOs are potentially dangerous, absolutely unnecessary, and a waste of resources. Only the massive power of the industry, an industry propped up by taxpayers' money and unable to attract private finance, prevents their being banned or simply quietly forgotten. Nevertheless, it was satisfying to see the long faces of the lobbyists from EuropaBio, the industry's mouthpiece, who, faced with what was essentially a defeat, appeared to retreat into a sort of adolescent 'Why does everyone pick on me?' depression.

On the other side of the fence, the bodies representing agricultural co-operatives and professionals working in agriculture, consumer groups and environmental lobbyists were strongly supportive of many aspects of the two Regulations, but raised the unanswered questions which were clearly not going to go away: the relationship between the agreed thresholds for contamination and the thorny issue of 'co-existence' between GM, conventional and organic agriculture, and the even more troublesome matter of liability, of who pays up if and when things go wrong.[54] Consumer groups were also unhappy about allowing adventitious contamination by non-authorised GMOs and wanted to see a commitment to further research 'into any unintended effects on human health or the environment or unintended changes in the composition of food'.[55]

Despite months of wrangling between the adoption of the Council's Common Positions on the two measures in November 2002, and the Second Reading vote at the European Parliament's Plenary on 2 July 2003, the Parliament substantially accepted the Common Position, which thus, with only minor adjustments, became law later in the year. The European Union finally had a set of laws governing the deliberate release of GMOs and the marketing of GMO products which at least represented something other than the corporate free-for-all which, as we shall see in the next chapter, rules the roost on the other side of the Atlantic.

### The United States' reaction

The US government had been beating the trade war drum for some time, with the British press reporting in August 2002 that it would bring a complaint to the WTO. This did not happen immediately, however, and for a number of reasons. Firstly, the United States did not want a trade war, especially as it was about to fight a real war and wanted European support. Secondly, it needed to maintain the support of Canada and others who shared its complaints about the labelling system,[56] but not its lack of concern for the norms of civilised discourse. Thirdly, the EU was already ignoring a longstanding WTO ruling that they must allow the import of beef from animals fed with Bovine Growth Hormone (BGH). The US had done nothing about this breach, partly because of its fear of sparking off a trade war and partly because it did not want to open a can of worms lurking at the bottom of which was a great deal of 'sound science' pointing to the risks to human health of BGH. Finally, the US was by no means guaranteed to win such a case. They had won a Pyrrhic victory over BGH, but there are significant differences between that case and any which might be brought in relation to the EU's GMO labelling rules. The EU had actually banned the import of beef from BGH-treated cattle, but in the case of GMOs all manufacturers and packagers are required to follow the same rules, so there should be no question of discrimination against importers.

Despite such considerations, the US did eventually take the European Union to the WTO, filing its complaint on May 13, 2003. 'The precautionary principle serves as an excuse for imposing arbitrary restrictions on new technology', argued two leading lights of US biotech in an article attacking the new EU regulations, while US trade representative Robert Zoellick, claiming that the United States had 'waited patiently for some four years' for the EU to drop its resistance, noted that the decisive factor had been the fact that 'intellectual dishonesty is spreading to other parts of the world' where some countries – Zoellick here accused China – were 'using it for blatantly protectionist purposes'.[57]

## SEEDS

Although seeds were included in the Regulations on food and feed and labelling and traceability, they were not covered by the thresholds established in those measures. Due to the vagaries of the EU legislative system, these thresholds had to be set by an entirely separate procedure and one, moreover, which excluded the European Parliament and the Council of Ministers.

The marketing of seeds within the EU, whether GM, conventional or organic is governed by Directive 98/95.[58] Only those seeds which receive the official status of 'variety' may legally be sold. GM seeds are required to undergo a specific risk assessment, but the Commission and member states have accepted that this is insufficient to ensure adequate protection of public health and the environment. As a result, in 2001 the Commission produced a new set of proposals in a working paper entitled *Adventitious Presence of GM Seeds in Seeds of Conventional Plant Varieties*.[59]

Under these proposals, contamination thresholds would be set at between 0.3 per cent and 0.7 per cent for approved varieties. No contamination would be accepted from unapproved GM material. Farmers would be forbidden from growing plants for seed use for a certain period after GMOs had been cultivated on the same land. Both the threshold for contamination and the precise length of this period (from two years up to five) would depend on the species of the seed, as would the stipulated isolation distance between GM and non-GM seed crops or crops in general.

Some of the proposals were embodied in a proposed Directive brought forward in 2002. However, industry lobbying had resulted in the loss of vital elements of the discussion paper's proposals. Gone were the statutory periods between cultivation of GM crops and crops for seed, whilst the isolation distances had been reduced. The Commission agreed to wait until after the adoption of the new legislation governing the deliberate release of GMOs and the marketing of GM food and feed before adopting the Directive on seeds, which at least gave the Council and Parliament the chance to influence it.

The Commission's view was that its proposed thresholds for contamination would be low enough to ensure that the products of such seed would not exceed the 0.9 per cent level, above which food and feed have to be labelled as GM. Contamination of seeds is by definition difficult to control, given that we are dealing with the means used by organisms to propagate themselves. Allowing any GMO contamination of conventional seeds could result in an uncontrolled release of fertile GM material into European farms. Recall of a GM product would be difficult; withdrawal of a GM seed, however, would be impossible. No farmer would any longer be able to decline to grow GM crops, no supermarket chain or food production company would be able to exclude GMOs. No consumer could refuse to eat them.

According to Commissioner for Consumer Affairs David Byrne, his proposals were 'geared to recognise the market reality that there is already unavoidable presence of GM material in certified seeds'. The reason for this is that 'Europe is heavily dependent on certified seed imports which are cultivated in areas of the world where conventional and transgenic cultivation co-exist to a significant extent.' In view of this, claimed Byrne, 'our proposal is designed to ensure that these levels are as low as achievable under the great majority of conditions'. The proposed 'tolerance labelling thresholds' – 0.3 per cent for rapeseed, 0.7 per cent for soya beans, and 0.5 per cent for everything else – would apply only to 'seeds for which a GMO authorisation for cultivation has been granted and for which the further use in food or feed has been authorised'. These thresholds were 'based on independent scientific advice' and were designed to 'ensure that the 0.9 per cent threshold set for food and feed is not exceeded'.[60]

The Commission submitted its proposal to the Standing Committee on Seeds in October 2003. Then the Directive would be submitted to the WTO for possible objections and could be formally adopted.[61] However, by September 2004, agreement had still not been reached, with the Commission, supported by a number of member states, sticking to its proposed contamination thresholds while other member states, most vociferously Denmark, held out for the 'detectability limit' of 1 per cent.

Frustrated, the Commission used its powers under Directive 98/95 to authorise 17 different seed strains of maize, all of them produced by Monsanto. The controversy was in no way reduced by the fact that the seeds already had national authorisations in France and Spain. Under 98/95, approval by one member state is approval to market the seed throughout the Union. Officially, provided all the rules have been followed, the Commission is obliged to extend that authorisation to an EU-wide basis. The debate over a new GM seeds directive, however, had been accompanied by a truce, and this power had never been applied. Unable to find an agreement, the Commission withdrew its proposal, returning, as it had long threatened to do, to an existing procedure. A revised proposal would be forthcoming, it promised, but in the meantime further delay to new approvals could no longer be justified.

## CO-EXISTENCE

Possibly the most urgent question left unanswered by the new legislative framework was that of so-called 'co-existence'. If GMOs were to be cultivated, what implications might this have for conventional or organic farms neighbouring those where transgenic crops were under cultivation? The EU line was that whilst there was no evidence that GMOs, if properly regulated, posed any danger to human health, people had a right to know what they were eating. Hence the labelling regime. Agriculture Commissioner Franz Fischler had acknowledged, moreover, that 'labelling will be worthless if we do not manage to segregate GM and GM-free on [sic] the fields of European farmers'.[62]

Yet the Commission had consistently shied away from tackling the problem, and resisted attempts by Karin Scheele to add obligations relating to co-existence to the Regulation on GM food and feed. Before the publication of the Commission's Recommendation on Coexistence in July 2003, the question of the regulation of GM crops within the EU was complicated by the fact that no one seemed to know where the right and responsibility to restrict or forbid particular cultivations lay or under what circumstances they might do so.[63] When the Welsh Assembly attempted to forbid the cultivation of T-25 maize – the scientific evidence supporting the application, and thus its approval under the old 'substantial equivalence' rules of the now-superseded Novel Foods Regulation having been entirely discredited[64] – the Commission told the UK government that such a move was not allowed under Article 16 of 2001/18. This states that while such decisions must endeavour to 'ensure a high level of safety to human health and the environment' they must also 'be based on the scientific evidence available on such safety and on the experience gained from the release of comparable GMOs'. According to the Commission, such evidence was in this case lacking. A similar decision prevented Austria from banning T-25.[65] Yet in an answer to a Parliamentary Question from Karin Scheele, and in the discussions on co-existence which preceded the agreement over the Regulation on Food and Feed, the Commission seemed to take the position that it would be up to the member states to decide the rules for co-existence, though the Commission retained responsibility to enforce the rights of those who wished to grow GMOs. This need not, of course, be contradictory, but the division between the two levels of responsibility was, to say the least, far from clear.[66]

Embarrassingly for the Commission, during 2002 the EU's own environmental research institute, the European Environment Agency (EEA), published a report which demonstrated that while some crops – potatoes, wheat and barley were cited, though potatoes raised for seed might be an exception – carried only a 'low risk' of contaminating others – certain others could be described as 'high risk'. These included the same rapeseed which, under its North American name, canola, had infested the Canadian prairies, and some fruit. In 'strawberry, apple, grapevine and plum...outcrossing and hybridisation tendencies...suggest that gene flow is likely to occur', while for 'raspberry, blackberry and blackcurrant' not enough was known for a prediction to be made. In between these high- and low-risk groups stood the 'medium- to high-risk' crops: sugar beet and maize.[67]

As well as the plant's reproductive techniques, upon which this list was based, the actual danger to the environment was affected by the presence or otherwise of wild relatives in the area, so that maize, which has no known wild relatives in Europe, was less of a risk than sugar beet, which does. On the other hand, this says nothing about the contamination of nearby crops. The report pointed out that 'none of these crops has pollen which can be properly contained' and recommended 'Management systems such as spatial and temporal isolation...to minimise direct gene flow between crops, and to minimise seed bank and volunteer populations' as well as 'isolation zones, crop barrier rows and other vegetation barriers between pollen source and recipient crops'. By such means pollen dispersal could be reduced 'although changing weather and environmental conditions mean that some long distance pollen dispersal will occur'. Just how much of a problem this might be is one of the many unknown factors which characterise genetically modified plants. 'The fitness of wild plant species containing introgressed genes from a GM crop will depend on many factors involving both the genes introgressed and the recipient ecosystem.'[68]

The report recommended a number of ways in which the problem could be addressed, including a review of isolation distances, even when crop plants had been engineered so that male plants were sterile, because study had shown that these 'will outcross with neighbouring fully fertile GM varieties at higher frequencies than previously thought'. Barrier crops might also reduce contamination, while 'Neighbouring farms should inform each other of their planting intentions in order for appropriate isolation measures to be considered.' Care should be taken to minimise 'volunteer and feral populations which act as gene polls carrying over the contamination into subsequent crops'. Finally, GM plants 'which incorporate biological methods to restrict the spread of transgenes between crops should be encouraged'.[69]

As for contamination of the broader environment by GMOs, it is clear from the report's conclusions that this is an extremely grey area which requires far more research. Factors about which not enough is known include the 'current levels of hybridisation and introgression occurring between conventional crops and wild species, and the behaviour of the new hybrids' which would 'determine the factors influencing the extent of gene flow and the likelihood of transgenes becoming established in wild populations'; '[the] geographical distribution of GM crop types and any wild species with which the crop is capable of hybridising'; 'the consequences of transferred genes in different species'; and '[the] stability of transgene expression over generations and in different genetic backgrounds'. On the basis of such research, and 'until we gain a better understanding' of these areas, 'Test protocols' should be developed 'to determine the likely effect of a transgene in a hybrid, so that on release of a GM crop the site can be surveyed for wild relatives and a risk assessment undertaken on a case-by-case basis'.[70]

The EEA's report was no isolated minority opinion in a sea of reassuring studies: evidence pointing towards the same conclusions had been accumulating for some time. In January, for instance, an official Danish working group had published a report concluding that not enough was known about a number of issues touching upon the question of possible co-existence, including '[the] importance of the extent of GM crop cultivation for the control measures to be adopted, [the] effect of buffer zones, [the] extent of cross-pollination, including the effect of field size, (conditions) affecting pollination, [the] potential for cross-pollination with wild relatives and volunteers' and 'the economic consequences of GM crop cultivation'. A study by British scientists the previous year had concluded that the popularity of organic farming, and the impossibility of having this side-by-side with GM, was rapidly reducing the number of areas where GM would be feasible.[71]

The title of the Commission's *Recommendation on Guidelines for the Development of National Strategies and Best Practices to Ensure the Co-existence of Genetically Modified Crops with Conventional and Organic*

*Farming* is self-explanatory. Decisions on co-existence would be left to the member states.

Or would they? As part of the final agreement on the two Regulations, an amendment had been inserted into 2001/18, explicitly granting member states the right to 'take appropriate measures to avoid the unintended presence of GMOs in other products'. Nowhere, however, was the term 'appropriate measures' defined. Nor were member states under any obligation to take such measures.

Giving member states total responsibility has two major drawbacks: firstly, it creates a lack of clarity, handing to the European Court of Justice (ECJ) the power to determine where a company's right to conduct its business ends and the member state's right to 'take appropriate measures' begins. The presumption behind EU law is that the right to trade freely within the single internal market may be abridged only under certain precisely defined conditions. Because of this, strong, clearly written EU-level laws are needed if restraints on trade, such as those necessary to ensure co-existence, are not to be open to challenge. The Commission's approach to co-existence most certainly fails this test. The second problem with this approach is that it paves the way for member states to compete to attract agricultural biotechnology, should they wish to do so, by indulging in a 'race to the bottom' in which the winner will be the country with the least restrictive co-existence rules.

These are not the only ways in which the system fails: nowhere is it clear who is to carry the cost of separating GM from conventional or organic crops, or pay up if co-existence arrangements are not adhered to, or if they fail. Member states do not have the right to declare GM free zones, as both the UK (in the form of the Welsh Assembly) and Austria (through five of its provincial governments) have attempted to do. In the Austrian case, the Commission decided that, although the Treaty's Article 95 does allow 'Member States to derogate from European Union harmonisation measures' this was only 'under certain strict conditions' which had not been met. The Commission's explanation of its decision referred to its *Recommendation on Co-existence*, which, it averred, 'states that priority should be given to management measures applicable on farm level and in close co-operation with neighbouring farms depending on crop and product type'. In other words, unless the Commission proposes a specific exemption from normal single internal market rules, no member state will in practice be able to declare such zones. Any company or individual prevented from growing GMOs who has complied with all the requirements of 2001/18 and subsequent legislation would be able to take the member state to the European Court of Justice, citing illegal restraint of economic activity. The contradiction here is clear: the whole of the legal framework governing GMOs is made up of EU-wide laws generated in Brussels and Strasbourg. Yet the Commission insists that each member state must take responsibility for co-existence rules.[72] The impasse threatened further to delay the lifting of the moratorium, with Austria insisting that 'Under present circumstances [we] cannot support the lifting of the moratorium. We need Europe-wide rules on co-existence first.'[73]

## LIABILITY

Another glaring omission from the EU's shiny new package of GMO legislation was the question, closely related to the problem of co-existence, of who should be liable in the event of something going wrong following a release. When 2001/18 was agreed, MEPs were persuaded to accept this by a promise that the matter would be dealt with in 'horizontal' legislation, which in plain English means a measure dealing with liability for environmental damage whatever the cause or source. When the relevant proposal arrived, however, it was hopelessly inadequate. Friends of the Earth, expressing the universal opprobrium of environmental NGOs, condemned the fact that its definition of biodiversity effectively meant that the Directive would apply to only 13 per cent of the EU's territory, that covered by the EU's Habitats and Wild Birds Directives and national measures establishing protected areas. This rendered it inapplicable to the most likely places where GMOs would be cultivated. Moreover, 'Exemptions are foreseen...that would let GMO producers and operators off the hook for any damage...', liability being 'precluded for any events or activities which have been authorised or which were not considered harmful based on scientific knowledge at the time'. Damage to neither property nor public health was covered in the proposal; no activities which took place before it came into force could lead to liability, and any action for compensation and redress must be undertaken within five years of the event to which it refers, a particularly unfortunate restriction in relation to GMOs, the long-term effects of which are yet to be seen. Nor did the proposal include an obligation on operators to take out insurance. As in so many other cases, the question of insurance is where reality tends to take over from rhetoric. The biotech industry can say what it likes: those who sell insurance for a living know well enough that GMOs are such an unknown quantity that offering liability insurance would be suicidal.[74] Finally, neither individuals nor associations may take action, the exclusive privilege of doing so being accorded to the member states in the form of their designated 'national competent authorities'.[75]

Attempts by some members of the European Parliament to improve the proposal were defeated, while the Council moved to water it down. Environmental groups were horrified and, in a letter to the Ministers, a number of organisations urged the removal of exemptions, a compulsory insurance system, and a broad definition of the ill-defined 'protected species' to which the draft text referred. Member states were divided, with the Greek minister, at the time in the Presidency of the Council, actually proposing a compulsory insurance system, addressing the most controversial issue by suggesting a maximum nine-year phase-in period, but the proposal won the backing of only four other member states and none of the big four, all of which stood firmly for a voluntary system.[76]

Member states were also divided over the breadth of permissible exemptions. The Commission insisted that no company should be prosecuted for an act which was either licensed by a public authority or in keeping with best known practice at the time it was committed. This would entirely exempt GMOs, apart from any which may be grown without authorisation, as unlicensed cultivation is illegal throughout the EU and licences are generally only granted to applicants proposing best possible techniques.[77]

The question of liability in the event of something going wrong is, in any case, only part of the issue. Arguably of equal if not greater significance is the issue of who should compensate the conventional or organic farmer if things take precisely the course expected in the best of all possible scenarios. This was highlighted a couple of months after the publication of the EEA's report on gene flow when a report touching in large part on the same subject was leaked to Greenpeace. Following the leak the report was at last published, but it had been delivered to the European Commission in January accompanied by a letter from the Director General of the EU's Joint Research Centre, Barry McSweeney, suggesting that 'given the sensitivity of the issue...the report be kept for internal use within the Commission only'. What McSweeney feared would be so controversial was that commercialisation of GM oilseed rape would increase costs for conventional and organic farmers by at least 10 per cent and possibly by as much as 41 per cent, with lower increases in the case of maize and potatoes. Co-existence would be in reality in many cases impossible, though this was already the obvious if unwritten conclusion of the EEA study. At best, massive changes in farming practices would be required, involving co-operation between all farmers in a region. As Greenpeace observed, 'it is not clear who would implement these changes, who would be responsible for controlling their correct implementation', or, perhaps the most controversial question of all, 'who would shoulder their costs'.[78]

Unfortunately, the industry's power is proof even against such clear evidence that strong action is needed. When the European Parliament held its second reading of the proposal, specific mention of GMOs was defeated. The one significant concession won was one which called on the Commission to propose within five years a harmonised compulsory financial guarantee for water and soil damage if no appropriate instruments or markets for insurance have been established. Species and natural habitats would then be covered in a further two years.[79]

## ENLARGEMENT OF THE EUROPEAN UNION

As the new package of legislation regulating GMOs was being hammered out, in the wider world beyond big things were also happening. One of the big things was the proposed enlargement of the European Union from 15 to 25 member states. In most areas of environmental legislation these aspiring EU countries had long been weaker than the existing member states, and they had a great deal of catching up to do. This meant that, on the one hand, activists hoped to be able to incorporate them into a relatively effective system of regulation and yet, on the other, feared that their entry might result in a weakening of controls.

A briefing paper produced in 1999 by the European Federation of Biotechnology observed that most Central and Eastern European (CEE) countries have no laws specific to agricultural biotechnology. There are, however, exceptions, and more have appeared since this was written. In some cases countries have introduced laws which may actually be weakened by EU accession. It is not clear, for example, that the EU approach to co-existence will allow Hungary to keep its Regulation 1/1999, which establishes rules for the creation of 'genetic protective zones' where GMOs would not be permitted. Poland even requires anyone releasing GMOs to take out insurance against environmental damage. Slovenia has a 'safeguard clause' which allows the national authorities to refuse approval to a GMO approved at EU level, whilst Slovakia's compensation law appears to be a model of the correct application of the precautionary and 'polluter pays' principles. In 2003 Croatia introduced a package of GMO-related laws which, while similar to those of the EU, were stronger in a number of areas.[80]

In principle new member states will have to embrace the EU's laws, strengthening or weakening their own as needs be. This will represent a huge improvement in, for example, Romania, where GMOs are so widely cultivated that the country represents what environmentalist organisations have dubbed a 'dumping ground for genetically engineered crops', but in other cases sound precautionary regulations may be weakened. Like Romania, Bulgaria has extensive GMO cultivation under regulatory conditions which would not conform to those now adopted by the EU. Yet both countries are officially due to join the Union in 2007. The ten countries which acceded in May 2004 have already agreed to adopt the EU system, but this leaves many questions unanswered, including what will happen about GMOs they have approved before accession, whether they have the will and capacity to implement these laws, and what the EU will do about it if they do not. Beyond this, if GMOs have been released which are not approved in the EU, have any provisions been made for the decontamination of the areas in which they were released?

Beyond the accession countries themselves, Russia and its neighbours have taken a cautious attitude to GMOs, though this may be changing. Russian legislation does not permit the development or production of GM foods. Such foods may, however, be imported for placing on the market. Officially they must be labelled, but in practice the corrupt and chaotic condition of the country's public administration means that this requirement is routinely ignored. Pressure to begin domestic cultivation comes from bigger farmers and not, interestingly, from the scientific establishment, which tends to favour a continued ban. Aggressive marketing by Monsanto, the weakness of any environmental or consumer protection movement, the lack of awareness of the issue amongst ordinary Russians, and the paucity of funding sources available to scientists are likely to erode resistance. Russia – in common with much of the Third World – risks becoming a dumping ground for unsaleable seeds and other products.[81]

## THE SIXTH RESEARCH AND DEVELOPMENT PROGRAMME

The European Community has been directing research funds into biotechnology since the beginning of its First Framework Research programme in 1984. Most has gone on agricultural biotech, with the emphasis being on the 'safety' of GMOs. From 1985 to 1999, 81 projects involving 400 research teams spent around €70 million in public money on trying to find out whether various GMOs were safe to grow and to place on the market. Other projects part-funded by the EU during that period included an investigation into the potential of genetically modified algae, engineering tomatoes to produce antioxidant carotenoids, and several which aimed at genetically modifying micro-organisms to produce new antibiotics. As well as research *per se*, the Commission gave financial support to attempts to develop a network of clinical centres working on umbilical cord blood stem cells, and to information-gathering exercises such as surveys of public attitudes to various aspects of biotechnology and studies of ethical and social aspects of biotechnology.[82]

The Commission's view is, however, that the commitment it has shown has not always been matched by the member states, and that this has had consequences for the private sector. In the Communication sent to the member states at Barcelona, it laments the fact that investment in research and development (R&D) in general lags behind that of the United States, a problem compounded by 'fragmentation of public...support, and...the low level of interregional cooperation'. This analysis guided the design of the EU's Sixth Community Framework Programme for research, technological development and demonstration activities, a five-year programme (2002–06) in which biotechnology would figure highly. The Communication mentioned 'Genomics and technology for health' as one of the Sixth Programme's 'thematic priorities', adding that the Commission would co-operate with the member states and the European Investment Fund (EIF) to 'develop a competitive bioinformatics infrastructure in support of biotechnology research and focus support for the development of research in computational biology and biomedical informatics'. Other initiatives would seek, amongst other things, to stimulate capacity for research: a 'strong, harmonised and affordable intellectual property protection system' would function 'as an incentive to R&D and innovation'; this would require member states urgently to transpose into their national laws 'Directive 98/44/EC on the legal protection of biotechnological inventions' and the adoption by the Council of an EU-wide 'Community Patent Regulation'. The Communication also outlines measures to improve the industry's capitalisation and its access to finance, to stimulate networks of businesses, scientists in public and private sectors and the universities, and regions, and encourage the adoption of good practices.[83]

#### Controversy over use of human embryonic stem cells

One area of controversy dogged the design of the programme and featured repeatedly in debates in the European Parliament and Council of Ministers. This concerned the use of human embryonic stem cells (ESCs). In some member states – the Republic of Ireland, France and Spain – this is wholly illegal. In most of the rest it is not prohibited, but researchers may use only 'supernumerary' embryos, embryos created as part of fertility treatment programmes which turn out to be surplus to requirements and, if not put to this use, would in any case be destroyed. In Denmark, Austria and Germany, these must be imported. Belgium, Italy, Luxembourg and Portugal have no legislation on their use. Only the UK specifically allows the creation of embryos for research.[84] Moreover, the law is in a state of flux, with Belgium, Denmark, France, Italy, Portugal, Spain and Sweden all considering new measures, while enlargement further complicates the matter. As the expert body responsible for advising the Commission on ethical aspects of scientific issues observed, 'the challenge of stem cell research' has meant that many countries have felt the need to introduce new legislation, and many are still in the process of doing so. Under these circumstances, during the period that the implementing provisions of the programme were being debated, the Commission agreed informally not to grant funds to any project involving human embryos or ESCs, unless these were already stored in banks or isolated in cultures. Although bitter differences of opinion, culture and tradition emerged during the debate,[85] the controversy revolved not only around the ethical issues, but also more general principles of EU law. Those who favoured allowing such research argued that, if something were legal in a certain member state, then it would be overstepping the Community's competence to exclude it from the possibility of funding. Those who opposed the use of human ESCs countered that the money for the programme came from all 15 member states and only if a procedure were legal in all member states should it be eligible. With Britain's industry benefiting from the uncertainties surrounding the use of human ESCs in the US, the UK was particularly concerned to defend what it saw as the rights of its scientists. The issue was resolved, though far from definitively, in November 2003 when the European Parliament failed to back moves to deny funding to research using human embryos, just as it had done when the Sixth Programme had been approved.[86]

### National and private research programmes

Notwithstanding the Commission's complaints about the paucity of private sector and national funding, the EU's own research budget represents only a tiny fraction of the overall publicly funded effort, which is itself dwarfed by investment by private industry. In 2003, the UK committed £55.5 million to proteomics and post-genomics research alone,[87] an increase of 40 per cent over the previous year described by *Nature Biotechnology* as 'part of a five-year, UKP246 million structural biology research program that will focus on protein structure and function modelling, bioinformatics, and regulation modification and expression analysis'. Other parts of the programme focused, significantly, on 'training...science PhDs for careers in industry by teaching them management and project planning skills', investigation of gene function in relation to neuroscience, and 'UKP40 million into funding stem cell research...in an attempt to take a world lead in commercial exploitation of the technology'.[88]

## PATENT LAW

A patent 'grants the inventor a limited period during which no one else can make commercial use of the invention without the permission of the patentee'. It is granted 'in return for disclosing the invention itself, so that after patent expiry, it can be used by anyone who so wishes'.[89] To be eligible for a patent, an invention has to be 'novel, involve an innovative step, be industrially applicable and...not contrary to morality'.[90]

Traditionally in Europe, living creatures have been excluded, as has anything which would more accurately be described as a discovery rather than an invention, though both of these exceptions have been eroded in recent times. Neither exclusion is clear cut, and the fact that there is scope for interpretation guarantees that interpretations will vary between different times and different countries. If living creatures are not patentable, for example, what about parts of them? If so, which parts, and under what conditions? Are genes, DNA sequences or proteins open to patent protection? Do we include micro-organisms? And when does a discovery become an invention? According to British patent expert Daniel Alexander, 'if a discovery has a technical aspect or makes a technical contribution, a patent may be granted'. However, 'identification of the boundary line between what is and is not a technical contribution...is difficult'. What it means in practice, he argues, is that 'the majority of human genetic inventions are likely to be directed to using genetic information in a practical way either to produce artificially human proteins, antigens or immunogens for the purpose of identifying certain antibodies or proteins or for producing research probes'. For this reason, he concludes that it would be 'difficult to think of a case of [a] commercially viable patent in this field which would not involve embedding the information in a product of this kind' so that the exclusion of discoveries is, as far as biotechnology goes, largely 'theoretical'.[91]

Patent laws are essentially national affairs, so in principle if the member states of the European Union wish to adopt different definitions of what is or is not patentable, this creates no problem for EU or international law. However, such variations can and do cause difficulties when it comes to trade. The European Union's attempt to create a single internal market thus makes the question of any possible variation in patent law between member states critical, and attempts to harmonise and rationalise patenting systems have been an important aspect of this process.

Patenting of living organisms, and above all the establishment of property rights on any part of the human body or products derived from it, arouses strong passions. Much of this is simply displaced anger, directed at biotechnology, or aspects of it, rather than at patenting *per se*. As bioethicist Nikolaus Thumm argues, 'the moral issues thrown up by regulations governing biotechnology patents are only a derivative of biotechnology specific problems [sic]. Patents are no more than rights regulating the ownership of biotechnological inventions.' This argument, however, can be taken too far. When Thumm pleads for all sides 'to treat the morality of biotechnology and the issue of patenting separately', he ignores the way in which the issue of patenting presents these moral aspects in a distilled form, making it natural that the patents office should become a battleground in the ongoing biotech wars. The relationship between patents and profits appears to many who are suspicious of biotechnology to cast light on the hypocrisy of an industry which constantly claims to want to feed the world and cure the sick.[92]

It is nevertheless important to remember that the granting of a patent does not imply anyone's approval of a process or product. All it means is that the patent holder has, until the date the patent expires, exclusive right to use of a product or process in the sense that, with exceptions recognised in national or international law, no one else may do so without his or her permission. It gives the patent holder no rights in the face, for example, of national laws banning a product.

### The European Patent Convention (EPC)

Most European countries, including every member state of the European Union, are signatories to the European Patent Convention (EPC), under which patents are issued through the European Patent Office (EPO). The EPC follows the usual rules which, with slight – but sometimes significant – variation in detail, now apply to patent law internationally. An invention is defined as such if it is new, involves an inventive step – in other words, some form of innovation – and is 'susceptible of industrial application'.[93]

Though an EPO-issued patent offers the considerable advantage of patent protection throughout most of the continent, it remains possible to apply for a patent at national level. The EPO is not an EU institution, and it does not, technically speaking, even issue a 'European patent', but a separate patent for each of its member states.[94] As things stand, there is no 'EU patent', though the Commission has issued a proposal for one which, not surprisingly, stands mired in controversy.[95]

The EPC does not provide detailed rules specific to biotechnological inventions. It does, however, exclude, in its Article 53(a) any 'inventions the publication or exploitation of which would be contrary to "ordre public" or morality'. Also excluded, in Article 53(b) are 'plant or animal varieties or essentially biological processes for the production of plants or animals'. This is wholly in keeping with patent law as it has developed historically, but, importantly, as we shall see in the next chapter, conflicts with the direction it has taken in the United States since the early 1980s. In the US, the Patent Office has refused any ethical decision-making role, arguing that this correctly belongs to Congress. There is therefore no equivalent to these rules.[96]

In Europe, however, aside from the fact that this exclusion 'does not apply to microbiological processes or the products thereof', that would appear to be that: for genetic engineers the game is up. Sure enough, when Harvard University applied for a patent on a genetically engineered mouse, the EPO turned it down. The oncomouse was a new variety, wasn't it? And therefore ineligible for patenting under the European Patent Convention? Unfortunately, things are not that simple.[97]

The EPO's ruling was overturned on appeal, which upheld Harvard's claim that the oncomouse was not a new variety but a new type of animal that could not be classed as a 'variety'. The Appeals Board agreed, but expressed concern regarding the ethical issues involved, especially animal welfare. Harvard's lawyers answered that the mice would be extremely useful in the battle against cancer and that this should be taken into account; that because the mice were 'super-susceptible' to cancer, fewer of them would be needed to test for carcinogens and therefore overall suffering would be reduced; and that the fact that they were intended for work in closed laboratories meant that they posed no threat to the wider mouse population or to the environment in general. These arguments, despite a further statutory consultation period in which third parties could send in reasoned objections, prevailed.[98]

In relation to medical biotechnology, legal authorities have not considered Article 53(b) as guaranteeing that human stem cells, for example, were excluded from patent protection. Article 52(a) makes a distinction between discovery of a new property or substance, which is not patentable, and the putting to practical use of that discovery, which in principle is. One example would be a natural substance which could be used as an antibiotic. Discover a micro-organism which has such a property and you will not be allowed to patent it. However, develop the process needed to transform the microbe into a usable drug and you can patent this process. In this case, you may be able to patent the micro-organism itself as part of that process. Identify a gene with a useful property and the same applies. Isolate and clone that gene, work out how its useful property might be exploited – how it is 'susceptible of industrial application' – and you may be able to patent the process, gene and all. This is why the game was not up for genetic engineers at all. They do not in fact take out patents on plants or animals, but on the processes of genetic transformation and the genes which form part of those processes. In the United States this is clearly allowed. In Europe, under the EPO, there was scope for interpretation and argument, offering employment to lawyers but few advantages to the rest of us.[99]

The first attempt to clear up some of the confusion by establishing a patent system applicable to all EU member states, was made in 1988,[100] but a decade was to pass before a Directive on biotechnological patents would be approved.

**Directive 98/44 on the legal protection of biotechnological inventions[101]**

The Directive was an attempt to clear up the confusion surrounding the thorny question of whether organisms and parts of organisms may be patented. It answers this question with a resounding 'yes', Article 3(1) stating categorically that the definition of patentability given in the European Patent Convention applies even in the case of 'a product consisting of or containing biological material or a process by means of which biological material is produced, processed or used'. Furthermore, Article 3(2) reinforces this, adding that 'biological material which is isolated from its natural environment or produced by means of a technical process may be the subject of an invention even if it previously occurred in nature'. The sole exception to this is also given in Article 3, which states that 'the human body, at the various stages of its formation and development, and the simple discovery of one of its elements, including the sequence or partial sequence of a gene, cannot constitute patentable inventions'. Here, however, we are back in the realm of words which do not mean quite what, to the lay reader, they may appear to mean, for Article 3(2) seriously qualifies this, allowing the patenting of 'an element isolated from the human body or otherwise produced by means of a technical process, including the sequence or partial sequence of a gene...even if the structure of that element is identical to that of a natural element'.[102]

Further clarification is offered by the 'Recitals' to the Directive, the series of statements which precede the actual legally binding text of every EU legislative measure. They are designed, for the most part, either to explain the problem addressed by the measure, to clarify its intentions, or justify aspects of the measure itself. Thus, while they do not have the legal force of the main body of the text, they may be referred to by national or EU courts or other authorities when disputes arise over the interpretation or application of that text. From the Recitals to the Directive on biotechnological inventions we can see that

> an element isolated from the human body or otherwise produced is not excluded from patentability since it is, for example, the result of technical

> processes used to identify, purify and classify it and to reproduce it outside the human body, techniques which human beings alone are capable of putting into practice and which nature is incapable of accomplishing by itself.[103]

Maggie Grace , a researcher who worked on the Directive with then Liberal Euro-MP Gijs de Vries, accurately described this argument as 'very odd', pointing out that it 'creates a patent regime for biotechnology inventions markedly different from [that for] other inventions'. Just because you have developed and patented a technique to extract a metal from its ore, you would hardly expect this to give you a patent which covered the metal itself, even if the pure metal does not exist in nature.[104]

Even after the Directive won the approval of the European Parliament and Council of Ministers, controversy continued. Following adoption, a Directive must be transposed by the member states, usually within a period specified in the text. In this case, however, the deadline for their doing so, 30 July 2000, came and went, a year later only Denmark, Finland, Ireland and Greece having passed the necessary legislation. The UK had passed parts of it and Italy, Germany, Luxembourg, the Netherlands, Austria and Portugal had submitted suitable legislative proposals to their national parliaments, where they were greeted with various degrees of resistance from a range of individuals and parties on the basis of left, green and religious motivations, producing some unusual alliances. In Luxembourg, the House of Deputies asked the government to call on the European Commission to renegotiate, citing 'ambiguities regarding the patentability of living matter'.[105] The Dutch government was divided over the Directive's requirements, with some ministers emphasising the growing importance of biotech to the country's economy,[106] while MPs from the biggest party in the coalition, the Labour Party (PvdA), joined opposition groups such as the Green Left and radical left Socialist Party, as well as fundamentalist Christians, in bringing pressure to bear on the government to challenge it.[107] This was successful insofar as the Netherlands did indeed take an action before the Court of Justice attempting to have the Directive annulled,[108] arguing that it had been brought forward on the wrong legal basis[109] and that it breached 'subsidiarity', the principle whereby the EU may only make law where there is some advantage of its doing so, as opposed to each member state's making its own laws. In addition, the submission alleged that the Directive was so vague that it breached the principle of legal certainty, that it went against certain obligations of the member states under international law, that it breached the fundamental right to human dignity and that the correct rules of procedure had not been followed in its adoption. The Court, however, disagreed on every single count and told the Netherlands to adopt it as soon as it could reasonably do so. Gradually, the missing countries presented their legislative proposals, but the Directive has still not been adopted in full by every member state. France, for example, has refused to adopt Article 5 on the basis that it is in conflict with a national law which forbids the patenting of human body parts.[110]

The case revealed a deep unease, in the Netherlands and beyond, over the idea of 'patenting life'. Underlying this sentiment, and the controversies and debates it generates, is one of the big questions which this book attempts to address: is biotechnology primarily a legitimate application of science, which can be developed to benefit all of us? Or is it – following the gradual sequestration into private hands of the world's productive capacity, and, over the last two decades, the accelerating privatisation of basic services and natural resources – the final usurpation, the privatisation of life itself?

Take the particularly sensitive issue of patenting of human DNA. As explained by a group of bioethicists writing in the prestigious journal *Nature Biotechnology*, 'Companies, joined increasingly by universities and research institutes, have seen patent ownership of human DNA as a potent means of gaining exclusivity in commercial markets for medicines, diagnostics, and research tools. Others have viewed such appropriation as unethical', partly for philosophical reasons but also because 'human DNA is composed of pre-existing information that has been discovered and not invented'.[111]

This objection, which could to some degree apply to any sequence from any genome, lies at the crux of the dispute over patenting. Behind it lies the broader question of whether patents help or hinder research in a particular field, and whether this should be the determining question. The European Federation of Pharmaceutical Industries and Associations' biotech lobbying arm expressed the majority opinion in the private sector when it wrote that 'Patentability of inventions fosters the development of scientific advance into benefits for patients and consumers...[and] will bring medical and therapeutic benefits.'[112] Publicly funded bodies tend, however, to disagree. In 1998 a number of research institutes and presiding authorities, including the European Commission, signed an international agreement to make any gene sequences they discovered immediately available. Secrecy, they argued, could only hamper vital research. This position has support in the private sector, with pharmaceutical companies disagreeing over what should and should not be patentable, and varying widely in the extent of access they are willing to give researchers in non-profit institutions. Those which do not do so tend to attract opprobrium from within the 'scientific community', with the *New Scientist*, for example, accusing biotech companies of 'gobbling up patents on everything from DNA sequences to altered animals' and 'driving too hard a bargain', warning that patents would 'accelerate research' only 'if they are pursued with the kind of spirit that fosters progress, rather than profit at any cost'.[113]

Those companies which pursue too aggressive and restrictive a patenting policy were accused by scientists of risking discrediting biotechnology as a whole, justifying its critics' tendency to ridicule claims to lofty aims such as 'feeding the world' or eradicating killer diseases. In 1996, the Consultative Group on International Agricultural Research, for example, warned that any chance of repeating the achievements of the 'Green Revolution' was being jeopardised, because 'patents held by companies will create a "scientific apartheid" which locks the 80 per cent of people in developing countries out of scientific advances'. One European patent owned by Monsanto covered any and every process of making transgenic beans, despite the fact that it described only one technique of doing so.[114]

Supporters of a patenting law sufficiently lax as to allow the patenting of discoveries tend to argue that patents stimulate research by giving researchers an incentive in the form of a likely financial reward. The contrary view is that a good patent system is one which balances the legitimate private interests of the researchers with the equally legitimate (or, for the left, prior) public interest in seeing scientific progress move as rapidly as possible in the direction of the general good. What those who hold this view fear is that access to health care services will be restricted by increased costs. In addition, the granting of inappropriate patents or abuse of existing patents may hinder the development of new medicines or treatments. More fundamentally, the free exchange of information, ideas and material which is the basis of scientific progress, and the openness which can help to ensure that such progress is at the service of humanity rather than in pointless or dangerous directions, can both be undermined if the patent system does not, while always ensuring that ingenious and hard-working people can profit from their work, give them priority. While patents may work to stimulate some forms of research, they would eliminate or restrict others. Citing the case of orphan drugs (those which, while proven to be effective in treating certain rare conditions, can find no manufacturer willing to produce them), an Italian Euro-MP, charged with preparing a report on the implications of human genetics, warned that patents would tend only to promote research aimed at developing 'diagnostic and therapeutic tools offering the prospect of substantial profit', while 'research would cease in fields that did not hold out the promise of the desired profit margins'.[115]

A further difficulty is created by the fact that the nature of an organism's genome means that it is unavoidable that patents will 'overlap'. Genes, as we have seen, have sections which 'code', known as exons, and those which have no apparent function, known as introns. A single gene may be covered by many patents: 'For example, a gene with 15 exons could well have a separate patent claim on each…another claim on the complete expressed sequence, a separate claim on the promoter sequence, and perhaps another on a distant locus control region.' Or 'each of a number of mutations of the same gene may be the subject of a separate patent claim'. The result of such a farrago of overlapping claims can be that, before a research programme can get under way, complex negotiations must be held with a large number of patent holders. One group of researchers had to acquire the rights to use of 39 separately patented sequences before they could begin their work, attempting to develop a vaccine against malaria. Moreover, a single patent may cover many sequences, with one filed by a German company claiming the right to 382,046 human DNA sequences.[116]

These continuing problems led to a legal impasse. By August, 2003, only seven member states had implemented the Directive to the Commission's satisfaction. The Commission took the other eight to court, and the verdict of the ECJ is awaited.[117]

### The Myriad patent

In addition to the ethical problems associated with biotechnology is the more concrete issue of cost. Granting patents on anything at all – be it a human gene or a novel form of swab – of importance in medicine is likely to increase the cost of health care. As in every member state of the EU health care is to one degree or another publicly funded, this goes beyond issues of consumer protection and involves the direct financial interests of the state itself. This can be clearly seen in the case of the controversy over the granting by the EPO of a patent on two breast cancer genes to a US firm, Myriad

Genetics. The patent, on a gene known as BRCA-1 gave them the exclusive right to screen for it and perform the necessary treatment. The company's persistent hindering of research by its refusal to license its patent out and its imposition of irksome conditions on those scientists who apply or are allowed to work on lines of research requiring access to the gene, resulted in what *Nature Biotechnology* described as 'ongoing opposition hearings at the European Patent Office'.[118] These hearings would lead, eventually, to the overturning of the patent.

Despite this, the controversy demonstrates that the exemption for 'non-commercial experimental or research use', which forbids patent holders to charge for such use, a sop to the opposition much trumpeted by industry lobbyists,[119] is of limited value, as it does not apply to non-profit use in general. Running tests on a non-profit basis probably seems a laughably altruistic activity to the folks at Myriad, who would no doubt be puzzled by the principles on which socialised health care is founded. Unfortunately, they are not the only ones. As early as May 2000, one NHS geneticist told the *New Scientist* that he had already had demands for royalties not only for breast cancer genes but for those associated with cystic fibrosis. The Directive offers no protection against those who see a sick person not as a suffering fellow human being but as a means to enhance profitability. Just as in gangster movies, protection is something that you buy.[120]

Opposition to the Myriad patent came not only from those people who object in principle to the granting of genes on the human body, but from others who, whatever their views on the ethical issue, saw the granting of such an exclusive right as both a hindrance to research and an invitation to profiteering. As one senior Dutch oncologist observed, 'I can't even analyse my own DNA any more without paying a great deal of money.'[121] Although the form of breast cancer involved constitutes only a small minority of total cases, in absolute terms this adds up to a lot of women. The test, which would be offered to women with a family history of breast cancer, can detect the presence of the gene before any cancerous symptoms appear, allowing preventative measures to be taken.[122]

Currently, any competent laboratory can screen for mutations on BRCA-1. The patent would have meant that only Myriad's own laboratories can do so. The initial test, which looks for four mutations, costs $40. However, further tests which, depending on these first results, may be needed, can cost as much as $9,000. The same tests could be carried out in Europe for less than a third of this.

Moreover, what really concerns researchers is that there is nothing exceptional about BRCA-1. It is only one of numerous genes which may be implicated in breast cancer susceptibility alone. The result of a wave of such patents would be that tests would become more expensive, but the problems do not end there. If laboratories holding patents insist on the exclusive right to carry out the related tests, not only is this a clear invitation to profiteering, it also means that the necessary expertise will wither away, leaving it concentrated in what is likely to be, given the realities of the market, an ever smaller number of private sector hands.

These views were broadly shared by public sector bodies in the EU and the people that work for them, as was shown when senior medical researchers and doctors from the Netherlands, France, Belgium, Denmark, Germany and the UK filed what is known as an 'Opposition request' – an objection to a patent – with the EPO, and the European Parliament by an overwhelming majority adopted a resolution condemning the Myriad patent and calling on the EPO – not for the first time – not to grant patents on the human body, or any part of it. The Resolution expressed concern for the social and economic impact of the granting of such patents – particularly increased costs to publicly funded health services – rather than what it perceived as an affront to human dignity, though this was mentioned during the debate. European Parliament resolutions have no legal force, and as mere expressions of opinion they are more likely to be influential if they are carried by a large majority and have support across the political spectrum. For this reason they tend to attempt to reflect a consensus rather than a simple majority view, and not all members would have been happy with a form of words which might appear to rule out each and every patent which could be described as a patent on the human body.[123]

The 1998 Directive does exclude from patenting any process aimed at modifying the germ line of human beings, anything involving the use of human embryos for industrial or commercial purposes, unless these may be useful to the embryo itself, and 'processes, the use of which offend [sic] against human dignity, such as processes to produce chimeras from germ cells, or [from] totipotent human or animal cells'.[124]

It also excludes from patentability any procedures for human reproductive cloning. This does not mean that all of the techniques which would be necessary to develop human reproductive cloning are also excluded. As any opponent of 'therapeutic cloning' will explain, the techniques necessary to any success in this field are precisely the same as those which would need to be developed before human beings could be cloned.[125] This is the major reason why the technique has been condemned not only by the religious right but by the European Parliament, in a resolution supported by Greens and some from the left.[126] This makes it less surprising that, according to expert commentators, 'A literal interpretation of the definition of human cloning in Directive 98/44/EEC would also include the initial step of the so-called somatic cell nuclear transfer technique (SCNT).' However, research in this area remains in its infancy and the fact that the 'development of SCNT has the important ethical connotation of unavoidably sharing part of its technical pathway with human reproductive cloning' means that 'the different treatment of "therapeutic" and "reproductive" cloning methods, if not carefully handled, could open the door for misinterpretations and carries the danger of opening the door to techniques to implement human reproductive cloning'.[127]

In February 2004, the EPO revoked one of the two patents involved, each of which covered a separate gene responsible for some forms of breast cancer. In May, the second patent was revoked, effectively ending the controversy, at least for the time being. Neither judgement, however, set a precedent which would prevent similar patents' being granted in the future: the first relied solely on the mundane discovery that the charity Cancer Research UK had in fact filed an earlier patent, precisely to prevent the possibility of commercial exploitation, while the second stemmed from inconsistencies between the DNA sequence described in Myriad's patent, issued in 2001, and the sequence in Myriad's original patent application seven years earlier. The original mistake was apparently corrected seven months after the application was filed, but by then, as the *New Scientist* explained, 'the crucial sequence had already been published openly elsewhere – so-called "prior art". That automatically made it unpatentable, because inventions have to be completely original and "inventive" to earn a patent.'

#### The Edinburgh patent

Confidence that the EPO can be trusted with such responsibilities seems misplaced, especially in view of its decision, in 1999, to grant the University of Edinburgh a patent which included human cloning methods. The patent was actually entitled 'Isolation, selection and propagation of animal transgenic stem cells', but in its text it was stated unambiguously that 'in the context of this invention, the term "animal cell" is intended to embrace all animal cells, especially of mammalian species, including human cells'.[128] The fact that in the face of the furore that followed the EPO claimed to have granted the patent due to linguistic confusion hardly improves the matter.

In fact, the controversial aspect of the patent should have been invalid, as under the European Patent Convention the two sections – the 'Description' of the patent and the 'Claims' made to back it up – must accord. In this case, though the definitions given in the Description did mention human cloning, no such process was actually included, invalidating its inclusion in the 'Claims' section.[129] Unfortunately, the matter did not end there, because the EPO is not empowered to amend a patent it has already granted unless someone asks it to. In its press release admitting its error, it solicited such requests or objections, known under patent law as 'oppositions'. They were not long in coming, with Greenpeace leading the way. Not surprisingly, given the position it had already taken on the issue, the European Parliament strongly condemned the patent, demanding moreover 'a review of the operations of the EPO to ensure that it becomes publicly accountable in the exercise of its functions' and that it 'amend its operating rules to provide for it revoking a patent on its own initiative'.[130]

However, four months after the patent was granted, the University of Edinburgh itself asked for the patent to be amended by the insertion of the words 'non-human' before 'animal' in two sections of the 'Claims'. Further confusion then set in, with the EPO shifting its ground. Apparently, it had now taken the trouble to read the patent properly and declared that it did not refer to cloning at all:

> Cloning is a process of asexual reproduction of an organism that creates multiple *identical* copies...of the original. This is achieved experimentally by nuclear transfer from a somatic cell into an enculeated oocyte. However, the patent...neither describes nor comprises nuclear transfer... the methodology described in the contested patent is not cloning.[131]

**Patents on products**

As well as processes, products can of course be patented. Most of the technology familiar to us in everyday life is subject to patent, or once was. Patents involving life forms, however, raise a new problem. The computer on which I am writing this is no doubt full of devices which are or were patented, but these devices lack the living organism's characteristic ability to reproduce themselves, a property which greatly complicates matters. The question might be summed up as follows: If you took out a patent on my mum, would that give you any rights over me? Fortunately, even the EPO has yet to start patenting people's parents. However, it would seem that, under the Directive, if a patent is granted on a process, it covers also the results of that process. As my mother has consistently refused to tell me anything about the process by which I was created, perhaps she intends to patent it, and were she to do so this would apparently give her a patent on me also. To be precise,

> the protection conferred by a patent on a process that enables a biological material to be produced possessing specific characteristics as a result of the invention shall extend to biological material directly obtained through that process and to any other biological material derived from the directly obtained biological material through propagation or multiplication in an identical or divergent form and possessing those same characteristics.[132]

A further problem is raised by the patentability of DNA sequences. If I discover that a particular sequence codes for the production of, say, a hormone which a particular plant uses to defend itself from a certain fungal pest, I can patent it. If, however, another team later discovers that this same sequence can be employed in the production of a molecule which would protect against a human fungal complaint such as thrush, they may not patent their application. Indeed, they could not work at all on the sequence unless they first paid a fee or otherwise persuaded the first team to let them do so. This provides further evidence that far from encouraging research by ensuring financial reward, such a system surely hinders scientific progress.

At the end of 2002 the European Commission finally admitted that new policy proposals were almost certainly necessary if the impasse were to be resolved. A group of experts in different fields was recruited to consider the various options. The Commission continued to insist, however, that the 1998 directive on patents did not need to be changed, and that everything was simply a matter of interpretation.[133] This attitude was criticised by those who saw a pressing need for reform. Sandy Thomas, director of the Nuffield Council on Bioethics, for example, whilst welcoming what he described as the Commission's acceptance that 'the mere claiming of a sequence where function has been guessed at through computer homology with similar sequences' should generally not be seen as sufficient grounds for the issuing of a patent, but wanted the European Patents Office to be obliged to 'tak[e] a lead in considering how the scope of claims to naturally occurring DNA sequences could be restricted to the uses referred to in the patent claims'.[134]

In response to these concerns, the European Commission published, in October 2002, a paper entitled *Development and Implications of Patent Law in the Field of Biotechnology and Genetic Engineering*. According to *Nature Biotechnology,* the commission 'admitted its thinking had been influenced by a report published in June by the UK's respected Nuffield Council on Bioethics', which argued 'that patents on gene sequences must become rare exceptions rather than the norm'.[135] The influence is clear, but the immediate motive behind the shift in Commission thought from that which was evident in the proposal which led to the 1998 Directive was pressure from the nine member states which had failed fully to implement its provisions.

The Nuffield Council, moreover, was expressing a growing consensus, rather than an innovative position. The *Financial Times*, for example, a newspaper generally sympathetic to the biotech industry's arguments, praised the Nuffield Council's report, noting that while acceptance of the 'principle that a gene sequence should be eligible for patent protection' may have been correct 'in the early days of DNA research', things had changed. In those times 'scientists had to make a heroic experimental effort to isolate an individual gene, understand its function and predict its utility' so it was understandable if 'patent offices took the view that the DNA sequences were not unpatentable natural phenomena but were created artificially by cloning genes away from the human body'. The replacement of physical cloning by computer-based methods of identifying genes made it much easier to 'stake out a claim to a stretch of DNA'. Characteristically, Craig Venter was even blunter, charging that companies were now able to file patents on DNA sequences 'without doing any work'. Since this is now the case 'only the most exceptional applications to patent a gene will satisfy the three legal tests: that it must be novel, inventive and useful'.[136]

The European industry, meanwhile, was concerned about the allegedly slow rate of biotechnology linked patent registration in the EU when compared to the US, and blamed the persisting confusion. Whilst it was true, however, that there was a growing gap between rate of registration of patents, it hardly seemed catastrophic. Comparing the period 1986–90 to the period 1996–2000, new biotechnology and genetic engineering patents rose by 239 per cent and 306 per cent respectively in the EU, as against 290 per cent and 344 per cent in the US, a gap which could surely be accounted for by the generally less friendly environment biotech had to cope with in the EU, rather than the specifics of patent law. As far as the particularly controversial business of patents on gene sequences was concerned, however, nobody seemed to know the precise figure, and the Nuffield Council was critical of the EPO for failing to collate such an important statistic. None of this stopped the *New Scientist* from describing the rise in the number of applications for such patents as a 'commercial stampede' which 'on both sides of the Atlantic' had 'Patent examiners...drowning in biotech applications'. Not only were there more patent applications, the average size of such applications was also increasing. One filed in 2000 contained over 140,000 pages, while the British Patent Office complained of an application which filled 'six big cardboard boxes'. Even those who defended the patenting of DNA sequences had to recognise that the system was buckling.[137]

It seems clear that, at the very least, patent criteria need to be applied much more stringently, so that patents based on research or diagnostic uses would generally not be granted, and compulsory licensing would be imposed where such patents exist and are hindering research or otherwise against the public interest. The Nuffield Council has suggested that a patent 'is defensible when the process of gene isolation translates directly into tangible products with a specific and readily identifiable use beyond their mere informational content'. In addition, because a range of proteins can, through the manipulation of different combinations of its fragments, be produced from a single gene, 'the rights to a DNA sequence should extend only to the production of the protein described in the claim'. For the utilitarian-minded Council, the crucial question is whether 'the benefits to society arising from strong incentives in the form of patents to produce novel medicines outweigh the disadvantages to others of restricting commercial use of a DNA sequence for the production of a therapeutic protein'. They are, moreover, not isolated in this view but appear to represent a new European consensus, with the European Commission citing their views with approval and the International Commission on Intellectual Property Rights arguing that patents on naturally-occurring DNA sequences should cover only the specific uses disclosed when the patent is filed.[138]

### Patents on plants

Apologists for agricultural biotechnology generally claim (quite falsely) that genetic engineering is essentially no different from traditional techniques of plant breeding, that it is merely a way of using recently acquired knowledge of biology to achieve greater precision. Yet they have chosen to use the patent system, rather than the alternative which has, in Europe at least, long been considered more suitable, the system known as Plant Variety Rights (PVR) or Plant Breeders' Rights. When this system was regularised internationally in the 1960s, plant and animal varieties, as well as systems for their production, were excluded from patentability. This was changed in 1991, when the International Union for the Protection of Plant Varieties (IUPpP: now the International Union for the Protection of New Plant Varieties (UPOV)) revised its regulatory Convention to allow either system to be used. This has, however, been ratified by neither the EU nor any of its member states, so within the Union it remains impossible to patent plants as such.[139]

Animal breeds, on the other hand, have traditionally no comparable legal system of protection. This makes it easy to understand why biotech firms are so keen to establish such a system, and patenting has the advantage of denying researchers the right to use protected varieties to research and develop new products. For this is the great boon that patents offer industry, and why PVR simply will not do. The PVR system contains an important exemption. Known as the 'breeder's privilege' or 'research exemption', it allows those seeking to research and develop new plant varieties to make full use of their predecessors' work. Without either seeking permission or paying a fee, they may use existing plant varieties in their breeding programmes, marketing any successful results commercially. In this respect the PVR system prioritises the public good, making the assumption that the more plant varieties are bred the better it will be for humanity as a whole. The patent system, though it takes the public good into account, represents a significant shift in the criteria upon which decisions are taken, offering greater control to the 'inventor' on the presumption that only by offering the incentive of such protection can the ingenuity and investment required to produce new plant (and animal) varieties be released. As things stand, however, neither system prevents farmers from reusing seed from a previous year's PVR-protected or patent-protected crop, as the 1998 Biotechnology Directive specifically safeguards this right.[140]

Nevertheless, transgenic plants do enjoy patent protection through the patenting of gene sequences, while in 2001 the EPO issued, for the first time, a patent on an animal, a transgenic salmon intended for human consumption. If you are wondering how this can be so, then you are not the only one to have done so. Shortly after the patent was issued, Jaime Valdivielso, a Spanish conservative Euro-MP, asked the European Commission if it felt that the patent was compatible with the ban on patents on animals. The Commission explained that, because the invention in question could be applied to other species of fish, and was not confined to a single species, it was in keeping with the directive. The patent on the salmon derived from the fact that it was the product of a patented process. In his question Valdivielso also referred to the risks involved in rearing such fish, should they escape into the natural environment. However, in response to this the Commission pointed out, quite correctly, that the issue of a patent did not imply any legal authorisation actually to farm the species, even experimentally, and referred the MEP to the *Directive on the Deliberate Release of Genetically Modified Organisms*, 2001/18.[141]

## FURTHER READING

Patrick Herman and Richard Kuper *Food for Thought: Towards a future for farming* (London: Pluto Press, 2003)

Andrew Rowell *Don't Worry, It's Safe to Eat: The true story of GM food, BSE and foot and mouth* (London: Earthscan, 2003)

International Centre for Genetic Engineering and Biotechnology *Biosafety Regulation in the European Union* <www.icgeb.org/~bsafesrv/bsfeurop.htm>

Monsanto UK *Facts on GMOs in the EU*, <http://monsanto.co.uk/news/2000/july2000/13072000gmoEU.html>

Gill Lacroix and Geert Ritsema *Conference Report: GMOs, co-existence or contamination?* (Friends of the Earth Europe et al., 2003)

Anne McLaren and Göran Hermerén, *Ethical Aspects of Human Stem Cell Research and Use* (European Group on Ethics in Science and the New Technologies to the European Commission, 2000)

Tony McGleenan *A Study on the Ethical Implications of Research Involving Human Embryos* (European Parliament Directorate General for Research Scientific and Technical Options Assessment, 2000)

Geertrui van Overwalle *Study on the Patenting of Inventions Related to Human Stem Cell Research* (European Group on Ethics in Science and New Technologies to the European Commission/Office for Official Publications of the European Communities, 2001)

Henk Jochemsen et al., *Human Stem Cells, Source of Hope and of Controversy: A study of the ethics of human stem cell research and the patenting of related inventions* (G.A. Lindeboom Institute and Business Ethics Center of Jerusalem, 2003)

Harald Schmidt *Harmonisation of EU Member States Legislation Concerning Embryo Research* (European Parliament Directorate General for Research – Directorate A – Scientific and Technological Options Assessment (STOA) Briefing Note No. 12, 2000)

Anne Eckstein (ed.) *The Patentability of Living Matter: Features from Europe Environment* (Europe Information Service, 2001)

Thomas Barlow 'Religious repression – western style', *Financial Times*, 8 December 2001

'Patents on life' (collection of articles) *New Internationalist*, September 2002, <www.newint.org/>

Friends of the Earth Europe Biotech Mailout (monthly), <www.foeeurope.org>

# 2
# The United States

## BOVINE SOMATROPIN

The first major impact which biotechnology had on the United States' food market was the introduction in 1994 of milk, cheese, butter, ice cream, yoghurt, beef and baby food containing bovine somatropin, or BST. BST is a genetically engineered hormone which also goes by the name of recombinant bovine growth hormone (rBGH). The Food and Drug Administration (FDA) approved the substance for administering to dairy cows – in order to stimulate higher milk production – without the slightest evidence that it was safe for long-term consumption by the people in whose food it would end up. There are fears associated with elevated levels of bacteria found in milk from BST cows, as well as pus and the residue of the high levels of antibiotic fed to these animals to counter their greater vulnerability to disease. The FDA was aware of these problems but simply chose to ignore them. It also rejected any requirement for labelling.[1]

Personal health was not the only reason why American consumers might have wished to avoid eating the products of cows injected with BST. There is considerable evidence that BST provokes mastitis, sores and numerous other health problems in cattle. BST also tends to exacerbate the overproduction of milk, leading to excessive consumption in a country where many people (especially African Americans) are lactose intolerant and problems of obesity, cardiovascular disease and other complaints associated with a diet too rich in animal fats are on the increase. Unhealthy levels of consumption are made possible only by government subsidy and massive publicly funded advertising campaigns, so that American taxpayers are footing the bill for propaganda designed to persuade them to eat more dairy products than is good for them.[2]

BST represents, moreover, a further intensifying of US agriculture, putting ever greater strain on the environment. There are fears, in particular, that it will lead to increased soil contamination and groundwater pollution.[3]

BST set the pattern for two of the most striking features of agricultural biotechnology: it provides a solution to a non-existent problem; and it would never have been possible at all without the input of huge amounts of public money. No doubt undernourished children would welcome the development of a product which can increase production of a rich source of protein and fat, calcium and an almost comprehensive range of micronutrients, by as much as a fifth. Unfortunately, this milk is destined not for starving Bangladeshi infants but for the already milk-saturated US market. The only reason farmers are interested in enhanced production is because the system of subsidies and price controls which prevails in the United States means that the more milk they can send through the farm gate, the more they earn, even though there is already far more than Americans can drink, more than they can consume in the form of cheese, butter, ice cream, yoghurt, and so on; more, arguably, than they could use if they all decided to start taking baths in it. Dairy farming in the United States operates on lines which are more akin to a distorted form of socialism than to anything which might be dubbed the 'free market'. If the Soviet Union had come up with such a system, it would now be used as a cautionary tale about the absurdities which result from a planned economy.

None of this would be possible without a huge corporate propaganda and intimidation machine, a pliant political system, and the systematic destruction of the independence of the American farmer. In that respect, BST was indeed a harbinger of what was to come, as biotechnology became the principal weapon of those corporations seeking to put the nation's (and the world's) food supply firmly in their iron grip.

## THE CORPORATE AGENDA

### The strange case of Drs Quist and Chapela

While the two scientists involved would have been aware of the controversial nature of their findings, the storm that hit Professors Quist and Chapela after the publication of their study in *Nature*, the world's leading peer-reviewed general science journal, was much fiercer than anything they ought reasonably to have anticipated. The two had made three mistakes: firstly, they had conducted a study which risked discovering that genetically modified maize had, despite a ban on its cultivation in part of Mexico known to be the plant's ancestral home, somehow contaminated farmers' fields containing criollo maize, a variety which also grows wild in that area; secondly, they had confirmed that this had indeed occurred; and thirdly, they had published their findings in a peer-reviewed journal.

Within hours, the paper was subject to detailed critique on a number of websites which drew attention to alleged flaws, as well as attacking the authors, especially Chapela. This turned out to be only the first wave of an immense, exaggerated reaction beginning with vilification from other scientists, anonymous abuse which turned out to come from a Monsanto hireling, and ostracism of Chapela at his workplace. One Berkeley colleague, Michael Freeling, demanded that since the two had 'published bad science in *Nature*, both scientists and *Nature* must come absolutely clean, retract and apologise'.[4] The bizarre pomposity of this statement was, moreover, by no means isolated. As plant biologist Lawrence Bush of Michigan State University later told the BBC,

> One of the surprising things about the reaction to Chapela's article in *Nature* was the vociferous and overwhelming disagreement that it conjured up. It was out of all proportion to the claims that were being made. In most instances in the sciences, if someone writes a paper that is seen as seriously flawed, one does one of two things. Either one produces some research that demonstrates how flawed that paper was, or one simply ignores it. An awful lot of work gets published that is flawed; it's just simply ignored.[5]

Next, the iron heel of the biotech industry came firmly down on the publisher. *Nature* received a mountain of mail, most of it hostile. The letters which were actually published, however, were from Chapela's colleagues at the University of California at Berkeley, or from others associated, through past employment or study, with the same institution. What concerned them can hardly have been the academic reputation of their employer or *alma mater*, for that had already been undermined by its acceptance of a grant of $25 million from biotech mega-corporation Novartis. Chapela had opposed the deal, as surely any scientist of integrity would, destroying as it did the independence of any research the beneficiaries might conduct and thus threatening the credibility of any findings they might publish. All of the letters to *Nature* were written by people who had benefited from the Novartis deal. Professors Chapela and Quist had apparently simply forgotten which century they were living in. They had to be made an example of, and the subtext of the faux indignation of the letters to *Nature*, as well as of the abuse they suffered in their place of work, was simple: get real or get out. The fact that the findings had been seized upon by anti-GMO campaigners was treated as proof that Quist and Chapela were guilty of giving aid and comfort to the enemy.[6]

Claims were made that the methodology, results and interpretation of their study were flawed.[7] The two researchers acknowledged that some of these criticisms were apposite, and specifically that they may have misinterpreted some of the results of their tests. However, few studies are entirely free of such errors: the point is to establish whether they are sufficiently serious to force a change in conclusion. The two insisted that they were not, that gene flow had in fact occurred and persisted, and that the transgenes had indeed located at several points on the genome of the contaminated plant. They even produced new data to back up their conclusions.[8]

Dr Chapela claims further that Mexican government officials tried both to intimidate and bribe him, offering him the chance to work 'for six weeks at a luxury resort'.[9] Even that, however, though crude, is not unprecedented: a scientist publishes a finding embarrassing or inconvenient to a government or rich corporation and is then offered inducements to retract or modify his conclusions. However, when *Nature* went so far as to state that its editors now regretted its publication because the evidence it presented had not justified its conclusion, it was at last clear that we were witnessing more than the usual spat between rival scientists. But then *Nature* had hugely overstepped the mark, for the propaganda machine developed by US agricultural biotech is so efficient, ruthless and unscrupulous that even a mildly sceptical comment in the mainstream or scientific press will result in a sort of 'shock and awe' overkill response.[10]

### The Monarch butterfly

This had already been seen in the case of the Monarch butterfly. As US readers will be aware, the Monarch holds a particular place in the affections of many Americans, and any threat to it from GM crops would lead to just the kind of publicity the industry would like to avoid.

The Monarch larva is one of those organisms which, because at one stage of its life it is entirely dependent on a single plant for all of its sustenance, is peculiarly vulnerable to any environmental disturbance. The larvae feed exclusively on milkweed leaves, so that, although vulnerable to the vagaries of nature, they have never made an enemy of the farmer, whose crops they leave unmunched.

Unfortunately, their kinship with less benign (from a human point of view) moths and butterflies, members of the taxonomic order *Lepidoptera*, undermines this advantage. The European corn borer, a major pest to US maize farmers and the main target of the toxins produced by transgenic maize, is of the same order. The question that Dr J.E. Losey and his team of researchers tackled was therefore whether the toxin could be killing Monarch butterflies.

Their conclusion was that this might indeed be the case. In laboratory tests, Monarch caterpillars fed on leaves dusted with Bt maize pollen ate less and grew more slowly than those fed on undusted leaves or leaves treated with pollen from non-GM maize. Moreover, within four days almost half of the 'Bt caterpillars' died while their 'conventional' brethren all remained healthy. The procedure was repeated several times and the results confirmed.[11] When the findings were reported in *Nature*, all hell broke loose, with anti-GM campaigners claiming that the study proved that Armageddon was upon us and the industry propaganda machine using its usual smokescreen techniques, describing the research as 'preliminary' (as if any of the scientists involved had claimed otherwise), and 'flawed', a catch-all insult that recurs frequently in the limited vocabulary favoured. With the collaboration of government agencies in Canada and the US, as well as agbiotech multinational corporations (MNCs), new studies were carried out. These established that the particular transgene used in Losey's study was indeed toxic to Monarchs and some other *Lepidoptera*, but that its use was restricted and decreasing. The most common transgenes used can be safely munched by butterflies and moths. In addition, Monarchs produce several generations in the course of a summer, whereas maize produces pollen for only around a fortnight, and due to such factors as rain washing pollen from leaves, caterpillars in the field would almost certainly get lower doses of any toxin present than did those used in the experiment.[12]

One small, fairly routine investigation whose findings did not suit the powerful forces behind agricultural biotechnology provoked not one further study but *six*. Each was a perfectly respectable piece of science which showed that, in this one instance, the dangers were not as great as Dr Losey's team's findings may have had us fear, and that where dangers did exist they could be minimised. What was lost – or, more accurately, successfully concealed behind a smog of verbiage – was what the findings certainly did show: once again, a technology which we have been told is no different to traditional crossbreeding and which therefore poses no significant or unacceptable risks to the environment or human health, was in fact capable of rendering a plant toxic.

The irony is that reactions such as that which greeted Losey's findings often rely on the fact that the vast majority of non-scientists, even amongst highly educated people, have little or no understanding of how science operates, or even of what it is. Thus, as in this case, the significance of mistakes in a study can be exaggerated or played down, or highly problematic concepts such as 'proof' and 'risk' can be manipulated to suit the industry's position, statistics massaged and tentative conclusions inflated into sensational findings. Coupled with the extreme bias of the US mainstream media towards the industry point of view, this does not make for balanced or truly informative coverage. A survey of major newspapers conducted in 2002, for example, found an overwhelming bias in favour of GM foods in both news reports and opinion pieces.[13]

Right wing pressure groups such as the Competitive Enterprise Institute do much of the bulldog work of routinely responding to any critical voice with a deluge of bluster supported by a thin sliver of distorted 'scientific fact', claims such as that Quist and Chapela's conclusions 'were later found to be wrong' and 'shown to be false' sprinkled invariably throughout.[14] Of course, anti-GMO groups are also capable of both exhibiting and exploiting scientific illiteracy, but the resources at their disposal are relatively limited. Even if Greenpeace were able to match the MNCs which dominate the industry dollar-for-dollar on publicity, they do not have a range of products to sell and a huge advertising budget to back them up.[15]

### Universities for sale or hire

As well as exerting immense influence on media coverage, the biotechnology industry and its agenda have now subverted US universities, undermining their independence. Attempt to pursue research which is not to the industry's taste, or publish conclusions which might be seen as adverse to its interests, and you do so at some peril. For every Quist or Chapela there are many who have simply decided that if they cannot go along with the pseudo-science promoted by the industry as the real thing, then they had best use their skills in another field. Ask a question such as 'Well, how do we know that GMO "x" does not have a negative impact on ecosystem "y"?' or 'What evidence do we actually have for the safety of this GM-based foodstuff?' and the next question you are likely to be asking is 'Would you like fries with this burger, sir?' Anyone who wants a career in plant biology in the US had better learn to love GMOs, and the evidence is that they have done so. As long ago as the early 1990s researchers were already noticing a change. One survey found that the idea of avoiding or declaring conflicts of interest, for example, no longer prevailed. Over a third of a surveyed total of 789 biomedical papers published in 1992 were written by people who stood to gain financially from the conclusions, and none of the authors involved mentioned this fact.[16]

Ten years later, another investigation found that 'Industry-funded research results in a much higher proportion of studies showing positive results for new drugs, compared to publicly-funded research.' Much of this research is carried out in universities. It can eventually result, for example, in drugs being marketed which have dangerous side-effects, or GM crops being grown which pose a threat to wild life. On a wider scale, however, the erosion of university independence could have consequences even more tragic than these, destroying the intellectual machinery which keeps the United States at the forefront of science and fields beyond it.[17]

## REGULATION

Events such as these help to explain why regulation of biotechnology and its products varies, in the United States, from weak to non-existent. This approach is possible, however, only because the regulatory authorities have been allowed to ignore the clear obligation, written into US law in the form of a 1958 amendment to the Food, Drug and Cosmetic Act, that any new food additives be tested for safety before they can be used in marketed products. The US government has simply found a way to evade its own laws. GMOs were not new: *ipso facto*, they required no such testing.[18]

For this reason, no specific agricultural biotechnology regime exists, and the regulation of the technology, its products, and the industries based upon them is handled by the same three federal government agencies as are responsible for comparable non-GM products. The FDA is responsible for food, animal feed, and pharmaceuticals. It decides whether a particular food is safe to eat or a drug safe to take. The United States Department of Agriculture (USDA) is concerned, as is obvious from its name, with farm animals and crop plants, their safety and suitability. However, a separate body, the Environmental Protection Agency (EPA), decides such things as whether and under what conditions a pesticide is safe to use.[19]

This multi-agency approach is reflected in political structures. In both houses of Congress, food biotechnology issues alone are covered by numerous committees, eleven in the Senate and five in the House of Representatives. The message is clear. As one Congressional research document puts it, 'The first wave of agricultural biotechnology food products is not substantially different from those foods already familiar and available to consumers.' Foods produced from GMOs will, the US authorities argue, be treated in the same way as any other foods: in other words they will be subjected to the same rigorous, exemplary safety procedures. The fact that much of the rest of the world does not accept this can be explained by the prevalence of superstition and ignorance. Whilst a nod is sometimes given to 'critics' who 'are concerned that this technology produces uncertainties about potential long-term impacts on public health and the environment', the real reason for foreign resistance to the marvels of US technology is simply good old-fashioned protectionism: it 'can be traced to their desire to allow their domestic industry time to develop a competitive position in this trade'.[20]

**Beginnings**

The first federal agency to become involved in the regulation of biotechnology was, however, the National Institutes of Health (NIH), which in 1976 published guidelines on research using recombinant DNA (rDNA) techniques.[21] Up to 1984, the NIH, through its Recombinant DNA Advisory Committee, was responsible for monitoring all such research, which at the time was moving from the realm of the speculative and into that of the commercially exploitable. In 1984, the Reagan Administration proposed replacing this by then inadequate arrangement with what it described, in a document of the same name, as a Co-ordinated Framework for Regulation of Biotechnology. This did not, however, represent an acknowledgement that biotechnological products needed any specific form of regulation. On the contrary, whilst it examined the problem of regulation it did so under the presumption that 'genetically engineered products would continue to be regulated according to their characteristics and not by their method of production', proposing 'that new biotechnology products be regulated under the existing web of federal statutory authority and regulation'. Because of this, the federal government saw its task as no more than identifying which agencies might most suitably concern themselves with such regulation. The Office of Science and Technology Policy (OSTP), an agency answerable to the

White House, was assigned the task of filling in the details of this broad approach and in 1986 produced a recommendation which was quickly implemented. As well as identifying which existing agencies would be most fitted to overseeing which aspect of the business of bringing GMOs from laboratory to field and from field to plate, it pre-empted possible turf wars by deciding who should be responsible where the division of labour was less self-evident.[22]

The Co-ordinated Framework for Regulation of Biotechnology, the structure adopted on the basis of the OSTP's recommendations, is substantially unchanged. It is not, however, the structure which is inadequate or inappropriate. The problem stems rather from the success of agricultural biotechnology corporations in influencing policy at the expense of the rights of the US citizen and consumer. One of the industry's greatest successes, for example, was to convince then Vice President Dan Quayle, in the early 1990s, that the regulation of biotechnology was 'the enemy of innovation – and therefore of economic survival'. The result is a regulatory system based on pre-existing legislation, one which focuses on individual products rather than the process of genetic engineering itself, which is seen as, in itself, insignificant. This can be seen in relation to each of the regulatory agencies, but is particularly stark when it comes to the FDA.[23]

### The Food and Drug Administration

The FDA is the body responsible for the regulation of foods on offer to the American consumer. Though it governs only foods traded between different states, in practice this gives it a broad remit to oversee quality and ensure the exclusion from the food supply of anything which may be injurious to health. It is charged with the inspection, for example, of food production facilities, as well as warehouses where imported foods are stored. In addition, it is supposed to control what may be added to food for purposes of preservation, flavour, visual appeal and so on. Because Americans buy most of their food in processed form, this is an important responsibility. Unless foods conform to legal requirements and are correctly labelled, the FDA may order their removal and, where appropriate, their destruction.[24]

This would appear to give the agency responsibility for the control of foods produced by modern biotechnology, and indeed it does. The problem comes in the weasel phrase 'substantial equivalence'. In 1992, under intense pressure from the industry, the FDA determined that genetically engineered foodstuffs or those in the production process of which GMOs had been used were in no way different in substance to traditional foods and therefore required no special regulation or labelling. Since then, new GM foods in general can be brought to market without any safety review.[25] As the *Washington Post* has complained, the FDA's 'approval' methods amount in legal terms to no more than 'a voluntary system under which biotech companies decide on their own how to test the safety of their products, submit summaries of their data...to the FDA, and win a letter that says, in so many words, that the agency has reviewed the company's conclusion that its new products are safe and has no further questions.' With few exceptions, none of the data needed to arrive at such a judgement are made public. The *Post* compared this unfavourably to 'FDA procedures for reviewing new drugs or food additives, in which the agency will spend months if not years going over company claims in detail'. In the case of GMO-based foodstuffs, the process from data submission to marketing approval can take as little as six months or, if you're really unlucky, as long as a year.[26]

There are exceptions to this, though they are inadequately policed. If, during the development of a new product, a gene transfer has produced unexpected results, the FDA is empowered to order a special review. If a naturally occurring species contains toxins, but in concentrations which do not endanger health, and an engineered variety has a greater concentration of these substances, the FDA may order a review, as it may if the levels of different nutrients differ markedly, if the added genetic material comes from a source known to provoke allergies, if antibiotic marker genes have been used in its production, raising the possibility that such genes might be taken up by pathogenic micro-organisms, lending them resistance, or if the product may have pharmaceutical as well as food uses. These restrictions apply equally to animal feed. Moreover, producers do now habitually consult the FDA before marketing new foodstuffs, including those containing GMOs or produced with them, though this is wholly voluntary and of benefit mainly to companies seeking to avoid future law suits.[27]

#### 'Generally recognised as safe'

Substances added to foods need no special regulation if they are classed, under the acronym GRAS, as 'generally recognised as safe'. Thus, for example, if we know that people have been eating carrots and apples for hundreds of years with no ill effects, and I wish to market a product which combines them, then the FDA will take no special interest. In cases of doubt, the FDA will require a producer to file an 'affirmation petition' containing evidence that a food is safe to eat. This apparently reasonable exemption has, however, been unreasonably and unscientifically stretched to include food products to which genes from other edible plants have been added.

The problem here is that when scientists introduce a gene into the genome of a plant they have no real way of knowing what the effects of that introduction will be, or whether the initial effects will remain. Genes are unstable, and their behaviour is poorly understood. Introduce the same gene into two different types of cells and you may produce two very different protein molecules. Introduce any gene and you will usually see significant changes in overall gene expression and therefore in the phenotype of the recipient cell and the organism of which it forms part. Moreover, as cell biologist David Schubert has warned,

> enzymic pathways introduced to synthesize small molecules, such as vitamins, could interact with endogenous pathways to produce novel molecules. The potential consequence of all these perturbations could be the biosynthesis of molecules that are toxic, allergenic or carcinogenic. And there is no *a priori* way of predicting the outcome.[28]

In simple terms, if you introduce 'machinery' capable of producing, *in situ*, the desired substances, these may spark off unforeseen, unwanted and possibly dangerous reactions resulting in the production of harmful molecules.

Yet the products of this hit-and-miss process are for the most part subject to no special regulation, though this appears to many critics to conflict with existing law. As activist lawyer Steven Druker has pointed out, 'The law specifies that in order to be classified as GRAS, foods containing new additives must meet two requirements; and GE foods do not fulfil either of them.' The first is a 'consensus in the community of recognised experts', which demonstrably does not exist. Even the FDA's own experts have warned about the possible disruptive effects of genetic engineering on the organism as a whole and the unpredictable possibility of the production of toxins. Secondly, the consensus must not rest 'on hypothesis but must be based on scientific evidence that clearly establishes safety'. It is clear that insufficient scientific evidence exists to make such a judgement, while a number of studies have pointed in the opposite direction. Indeed, as Druker notes, 'because the FDA does not require tests of

GE foods and performs no substantial scientific reviews, it still has no evidence demonstrating that any are safe'.[29]

The 'substantial equivalence' judgement according to which all GM products currently available in US food stores were approved is arrived at by measuring components such as sugars, proteins and micronutrients to see if these are roughly the same as those found in the parent organism, or foods produced from it. Safety tests on animals or human beings are considered unnecessary, even in the case of groups generally considered vulnerable, such as pregnant women, nursing mothers, babies, children or people who, for reasons of culture or poverty, are likely to have an unusually high proportion of a certain food in their diet.

Although the FDA is supposed to be notified and to act in cases where unexpected results are recorded following a genetic intervention, even if this system were fully enforced – which seems extremely unlikely, given the huge amounts of money at stake and the considerable interest of researchers and their employers in concealing negative results – it would not be adequate. In effect, consumers are being used as guinea pigs in trials of new foodstuffs on which no real testing has been carried out. The word 'unexpected' means that changes may be considerable and detectable without the FDA's feeling the need to take action. Calgene's Flaw Saw tomato, for example, was the subject of a successful petition to the FDA in 1992–94, shortly after the FDA had decided not to institute special regulation of GM foods deemed 'substantially equivalent' or where the added genetic material came from a source defined as GRAS. Although Flaw Saw contained a non-tomato protein, and the FDA originally classified this as an additive which would therefore need prior marketing approval, Calgene petitioned to have this decision reversed.

Unfortunately for Calgene, and in contradiction of the myth of the eager and trusting American consumer, the product was a flop. Campbell's Soup, who had financed the research which led to Flaw Saw, declined to add the new tomato to their products, while attempts to find alternative uses ended in failure. The debacle led in effect to Calgene's demise as an independent firm, as following poor financial results they were taken over by Monsanto. The irony is that the Flavr Savr GM tomato was at the same time successfully marketed in the UK in the form of a purée, despite being duly labelled as containing GMOs. It tasted good and was cheap, and was therefore a success, though when public interest groups began to express their concerns about the technology and sales fell, it was quietly withdrawn. Unlike in Europe, however, the Flavr Savr, which arrived on supermarket shelves in California and Chicago in May, 1994, was the beginning of a swift transformation of food, so that by the end of the century it would be difficult to avoid buying GM products. Indeed, without a radical change in diet, it would, for most Americans, be impossible.[30]

### No labels

It is this insistence on the 'substantial equivalence' of GM foods that has created the controversy over the US failure to require that such products be labelled. As long as the food is deemed 'safe', the consumer has no right to information regarding the methods through which it was produced. Even if an engineered food has failed to win approval as 'substantially equivalent', food manufacturers are careful not to give more information than strictly necessary to conform with the law, and particularly to avoid setting the precedent of mentioning genetic engineering, biotechnology or GMOs. In the case of a rapeseed (or canola) oil whose composition had been changed by genetic engineering, for example, the new product is labelled simply 'high laurate canola', a description of little or no use to the vast majority of consumers. Even the right of producers to label their foods as 'GM-free' is frequently called into question, despite a clear 1997 court ruling that state laws preventing this (they actually referred to rBGH only, but the principle is the same) represented an unconstitutional violation of First Amendment free speech rights. The question again arises, if US corporations are so keen on genetic engineering, why are they opposed to labelling?[31]

There were signs towards the end of the Clinton Administration of a certain softening of the stance against labelling. Promising to commission an independent review of USDA procedures for approving crops, Agriculture Secretary Dan Glickman stated the view that 'some type of information labelling is likely to happen'. Unfortunately, Glickman was talking outside his remit, because it is the FDA and not the USDA which would be responsible for such a system. He may have been making placatory noises as part of an attempt to stave off a trade war with the EU. Or he may, as Michael Meacher would later be when sacked as UK Environment Minister for his sceptical views on GMOs, simply have been an isolated voice in a generally pro-biotech government. In any case, he and his president were shortly after that out of office, and biotechnology's place in the sun was assured by the presence in Washington of its very best friends.[32]

Around the same time, however, one of the United States' largest food processors, ADM, announced that it would begin requiring suppliers to separate GM and non-GM crops. Insisting that it had not changed its view that GM foodstuffs were perfectly safe, ADM claimed, plausibly enough, merely to be responding to one of the normal pressures of the market, consumer choice. 'Some of our customers are requesting and making their purchases based on genetic origins of the crops used to manufacture their products', its statement to farmers explained, noting that 'If we are unable to satisfy their requests, they do have alternative sources for their ingredients.'[33]

There have been signs that the problem may extend beyond the European Union, and that other countries might begin to follow the EU's lead in demanding the right to control imports of GM products as a special category of commodity, a stance which clearly requires a labelling system. In March, 2002, for example, a ban on GM imports into China was only headed off after last-minute diplomatic manoeuvring. The following year, a number of African countries refused to accept GMOs in food aid.[34]

When groups of concerned citizens succeeded in having a proposal to introduce mandatory labelling of GM products added to a ballot in Oregon in 2002, the industry poured vast sums of money into the state. Biotech corporations established a war chest of $6 million, while pro-labelling groups expected to be able to raise perhaps $150,000. Not only that, but the federal government waded in on the side of the industry, the FDA warning that the measure 'would impermissibly interfere with manufacturers' ability to market their products on a nationwide basis'.[35] The proposal went further than the measures adopted in 2003 by the EU, as it would have required the labelling of foods made with enzymes produced from GMOs, and meat and dairy products from animals fed on GMOs, two categories excluded from the EU labelling requirements. In addition, the threshold for permitted contamination was 0.1 per cent, a level which had been rejected by the European Parliament in favour of 0.5 per cent, which itself had proved too low for the Council of Ministers to swallow. The industry, however, did not concentrate its fire on these features, presumably because to have done so might have been to appear to admit that labelling in itself would not be unacceptable. Organising themselves as a bogus citizens' movement, the Coalition Against the Costly Labeling Law, industry propagandists pumped out a series of lies, mostly on the theme of expense. Much of their campaign revolved around the claim that the measure would cost the average Oregon family $550 per year, a figure based on a report by a 'consultant' hired by the industry, and bearing no relation to the experience of any of the countries where labelling was already required. Even if the measure had been approved, it might well have been defined as an unreasonable restraint of trade by federal courts, falling foul of laws guaranteeing freedom of interstate commerce.[36]

### The FDA and the food industry

The weak regulation of GM food safety and labelling clearly reflects the power of the industry and the influence it is able to exert over the FDA. Movement of personnel between the FDA and big companies such as Monsanto is routine, so much so that the agency has become little more than an instrument for self-regulation, or what one consumer activist labelled the 'Washington branch office' of the biotech food industry. Behind this, moreover, lie close business and personal ties between members of successive administrations and leading lights in the biotech industry, as well as major contributions from Monsanto and others to political candidates. Bush's Attorney General John Ashcroft, for example, had been the biggest recipient of Monsanto funds during recent elections to the US Senate. Donald Rumsfeld, Bush's Defense Secretary, had previously been president of Searle Pharmaceuticals, now owned by Monsanto. Agriculture Secretary Ann Veneman had been on the board of Calgene Pharmaceuticals, another member of the Monsanto empire. Clarence Thomas, appointed by Bush Sr, and the man whose vote on the Supreme Court secured Bush Jr the presidency, was Monsanto's top lawyer. Larry Combest, the Texan Republican who chairs the House of Representatives Agriculture Committee, was the second biggest beneficiary of Monsanto generosity during the 2000 election. Richard Pombo, chair of the same body's Sub-committee on Dairy, Livestock and Poultry, received financial support from Monsanto in 1994 while working to defeat a bill to make labelling mandatory for milk products containing rBGH. Only days before ProdiGene's negligence almost led to pharmaceuticals entering the food supply, Bush had appointed its Chief Executive Officer to the federal government's Board on International Food and Agriculture Development. And so on.[37]

### The United States Department of Agriculture

USDA has the power, derived from two pieces of legislation, to regulate the movement of any plant which may pose a threat to agriculture. Under these two acts – the Federal Plant Pest Act and

Federal Plant Quarantine Act – new regulations were introduced which dealt specifically with genetically modified organisms. Not all GMOs are subjected to tests to see whether they may pose a threat, however, as once again wide exemptions are allowed for.[38]

The agricultural counterpart to 'substantial equivalence' is known as 'familiarity'. Familiarity is determined on the basis of whether the GM plant can be compared to its non-GM parent plant or counterpart, taking into account the biology of the species involved, the trait or traits introduced, and the broader agricultural system of which the GMO will form part as well as any possible interactions with the environment. Familiarity may also be invoked where a GMO has previously been approved for release. As with substantial equivalence, familiarity provides a broad waiver on treating GMOs as anything other than business as usual, what Ignacio Chapela has called 'a very anti-scientific principle of not looking', of assuming 'that these things, which we know are being produced through technologies that we never had access to before, are familiar...that they are not different'. With this 'mandate not to look' GMOs can be developed without being subject to the tiresome and profit-unfriendly business of safety tests or environmental assessments.[39]

### The Animal and Plant Health Inspection Service

The Animal and Plant Health Inspection Service (APHIS) of the United States Department of Agriculture is the federal government agency charged with governing imports and interstate trade in living organisms with the aim of preventing the spread of pests, though in the case of transgenic animals, a separate USDA agency, the Food Safety and Inspection Service (FSIS), has the responsibility to ensure that they do not pose a threat to human health. APHIS has the power to carry out an environmental assessment, to evaluate, for example, the likely environmental impact of the release of a novel crop plant or other organism. Only if the result of its assessment is FONSI – 'finding of no significant impact' – does it issue a permit, and then only after consultation with the authorities of the state or states affected.[40]

Sounds reassuring, except that in practice none of this applies to genetically modified crop plants. In 1993, at the same time as the federal authorities were deciding not to require GMOs in food or feed to be mentioned on labels, APHIS ensured consistency by deciding that what was the same in the can or freezer cabinet must also have been the same in the field. Initially, the agency approved six trials of GM crops without any environmental assessment having been carried out. Since then, any company wishing to grow GMOs need do no more than notify APHIS and complete a greatly shortened and simplified application form to which APHIS has 30 days to reply. Since the procedure was expanded in 1997 to cover almost all GM crops, only around one in a hundred applications must be made through the old regulatory procedure. The result is that new GMOs are constantly being introduced into the environment in the United States with no prior assessment of environmental effects. Only when test trials are completed and an application is submitted for the cultivation of the GMO with no special restrictions is an environmental assessment carried out.[41]

GMOs which fail the test of 'familiarity' are assessed case by case. The criteria are much the same as those used in the EU: is the GMO likely to become an infesting weed, or an invasive threat to the environment? Is it likely to pass its genes on to wild relatives or neighbouring crop plants? How might it interact with wildlife or the broader environment? The difference is that they are rarely posed.

### The Environmental Protection Agency

The EPA is responsible, under the Federal Insecticide, Fungicide and Rodenticide Act (FIFRA), for regulating chemical inputs to US farms. This has a direct effect in the case of GMO plants which have been engineered to produce their own insecticides. Because the breeding of such plants is (ostensibly, at least) aimed at reducing such inputs, the EPA is responsible for their regulation. In permitting a GMO of this type, it must decide whether the substance it expresses might cause 'unreasonable adverse effects on the environment'. In doing so it is also permitted to take into account any effects on overall use of pesticide or herbicide, balancing these, where they are judged positive, against any adverse effects. Before 1994, all releases of GMOs needed EPA approval. However, a relaxation of the rules in that year means that EPA approval is no longer needed if the gene alteration in question is deemed one which could occur in nature, such as the deletion rather than the addition of a gene. In order to make such evaluations, small scale field trials on a maximum of ten acres, approximately four hectares, regulated not by the EPA but by the USDA's Biotechnology, Biologics and Environmental Protection (BBEP) unit, are held, after which full-scale EPA-approved trials may be proceeded with (through the issue of an Experimental Use Permit – EUP). There is, however, no follow up monitoring of crops after commercial release has been authorised, and no attempt to study the possible long-term impact.[42]

In the case of food crops, the EPA must determine a 'safe level' of pesticide residue, one at which there is 'a reasonable certainty that no harm will result from aggregate exposure to the pesticide chemical residue'. In assessing this level the EPA takes into account 'all anticipated dietary exposures and all other exposures for which there is reliable information'.[43]

This form of words seems chosen to allow any size of vehicle to be driven through its loopholes. Firstly, if a dietary exposure is not 'anticipated' then the EPA has fulfilled its responsibilities and the producer in question complied with the law, even if some individual consumers drop dead, go blind or begin to believe that they can fly. Individuals, through poverty, personal eccentricity (of a kind which is typical of children, one of the more vulnerable groups) cultural or other factors, may consume far greater levels of a particular foodstuff than 'anticipated'. Moreover, if there is no 'reliable information' about a pesticide, it would seem that you can soak people's breakfasts in the stuff and enjoy complete immunity, a 'suck it and see' principle rather than the precautionary principle now written into, for example, basic EU law.

Nevertheless, these laws do provide some protection for the consumer and the environment, however inadequate this may be. The EPA has, moreover, shown itself willing to act in extreme cases, as when it refused to authorise a GM cotton variety engineered to be used with bromoxymil, a herbicide which has been shown to cause cancer and birth defects. Surprise, surprise: the rules do not, in practice, apply to substances produced by plants themselves for their own protection. As a 1999 research report to Congress put it, 'Because no tests of the registered [approved] plant-pesticides have shown toxicity to humans so far, EPA has given them an exemption from the requirement of a tolerance level.' The same report states confidently that there is nothing to worry about because the approved plant pesticides are, in all but one case, toxins produced by *Bacillus thuringiensis*, and that these are 'species-specific, affecting only certain insects. They are also virtually harmless to humans and animals.'[44]

## REGULATION AND PUBLIC OPINION

Americans are routinely said, without any evidence being presented to back the assertion up, to 'trust' the FDA and other federal regulatory agencies. If this is the case, then the arrogance and lack of interest in public concerns shown by both the industry and the officials who appear to be in its pocket may be eroding more than the public's willingness to accept GM foods. Growing disquiet, focused not so much on GM crops, animals and foods themselves but rather on the way in which they have been introduced into the food chain in the face of what seems like a complete failure of the regulatory authorities to exercise their responsibilities to protect the US consumer, public health and the environment, is undermining confidence in their willingness and ability to do so. Mainstream consumer groups, for example, have condemned the US authorities' lax approach, joining EU organisations in the Trans Atlantic Consumer Dialogue (TACD) in welcoming the package of proposals which led to the system of labelling and traceability discussed in the previous chapter by stating that such a system 'comes high on TACD's agenda' and that 'an alternative to GM must be available to consumers'.[45]

All of this fuelled growing scepticism in a country which we are constantly told welcomed GMOs with open mouths and in which, according to one survey, 60 per cent of the population had no idea that they were eating GMOs. The months following the Monarch revelations saw the first direct action against GM crops of the kind which had by then become commonplace in Britain. This was, moreover, accompanied by concerted legal action. In December 1999 a lawsuit was filed in numerous countries simultaneously alleging that companies have too much control over decisions taken by farmers over what to grow and how to grow it. Legal challenges were also made to the approval of plants containing the Bt toxin.[46]

People's desire for clear information about what they are eating is apparently to be dismissed as 'unscientific'. There are two problems with this. Ask someone who follows the food rules of Islam, Judaism or any other religion whether they consider those rules 'scientific' and the only honest answer that they can give is that the question is irrelevant. There are many reasons to want to avoid eating this or that food, and only some of them relate to anything which might be termed 'scientific'. Jews and Muslims must avoid a long and complicated list of foods and combinations of foods if they are to respect the rules laid down in their holy books. Most British people are averse to eating horses and snails, which are consumed in great numbers just over the water in France and Belgium. The Japanese in general find the very idea of cheese revolting. I have personally provoked disbelief in a US restaurant by asking for vinegar to put on my 'fries', whereas in my home city of Manchester it would be regarded as somewhat eccentric to eat chips without. Aversions and attractions are part of the diversity of human culture, and respect for them is generally regarded – at least outside food industry circles – as a good thing, a mark of a civilised and humane mind.[47]

People may wish to avoid GM foods for political reasons, believing, for example, that agbiotech will reinforce corporate control of the food supply. Political motives for seeking to avoid buying certain foods are also, generally, regarded as a normal and acceptable way to express dissent. Nobody pretends that there is anything especially unhealthy about foods from Burma, Libya, Iraq, France or any of the other countries which have been the target of boycotts or attempted boycotts in recent times. No one calls for a boycott of Nestlé on the grounds that its chocolate contains endocrine disrupting chemicals, and rum-topers don't seek out Havana Club instead of Bacardi (or vice versa) because they believe the latter is more damaging to the liver.

Environmental concerns may also lead consumers to wish to discourage the spread of GM crops by voting with their shopping baskets. Such action has led to major British supermarkets – including Asda, which is owned by Wal-Mart, America's biggest retailer – insisting on GM-free supplies.[48]

Why is all of this taken for granted in Europe and yet regarded as an assault on all that is held dear on the other side of the ocean? American consumers are beginning to ask just that question, and it is clear that the American biotech industry's attempt to discredit EU regulatory systems has not only failed, it has backfired. Many Americans can by now see as clearly as anyone else that whereas regulation in Europe is far from perfect, their own is, in any real terms, non-existent.

### A weak regulatory environment

The FDA relies entirely on the producers themselves for safety data, setting up a clear conflict of interests which may result in lax standards of enforcement. It conducts no tests of its own but relies on a review of studies paid for by the industry. Tests have not dealt with possible chronic effects, which are clearly at issue when a novel foodstuff may be fed to weaning babies and eaten at every stage of life thereafter. Few scientists not working for, or funded by, the biotech industry have been able to study data possessed by the industry on such crucial matters as what genes have been introduced into which plants, the genetic engineering methods used, or gene expression and how it is affected by a variety of conditions. According to Charles Benbrook, scientists independent of the industry have never been given access to these data, nor have any 'received the funding, information, and technical cooperation required to carry out what any team of experts would consider a thorough and independent assessment of GM safety claims'. The General Accounting Office, the Congressional agency which oversees the work of federal bodies such as the FDA, has criticised it for not examining test data with sufficient thoroughness.[49]

#### Signs of change

Despite the entrenched power of the industry, the disappointing result of the Oregon ballot and the Bush Administration's unqualified support for agbiotech, there are signs that US consumers are waking up to the giant trick that has been played on them. One example of this comes from a survey which supposedly showed that 'American support for food biotechnology is holding steady'. The survey, paid for by the International Food Information Council (IFIC), an industry front body whose members include Monsanto, Aventis, ADM, Cargill and other major agbiotech players,[50] was subjected to critical analysis by the public interest group PR Watch. The group concluded that it was 'so biased with leading questions favoring responses that any results are meaningless'. Typical questions were along the lines of '... how likely would you be to buy a variety of produce, like tomatoes or potatoes, if it had been modified by biotechnology to taste fresher or better?' and 'If cooking oil with reduced saturated fat...was available, what effect would the use of biotechnology have on your decision to buy this cooking oil.' The survey did not actually ask consumers whether they would buy a GM product which would make you more sexually attractive or richer, or guarantee a pain-free lifespan of 200 years, but it might as well have. No mention was made of possible concerns about health or the environment. Buried in a list of somewhat lukewarm endorsements of biotech's wonders, moreover, we discover that only one in five US citizens believes the industry-generated myth that GMOs will reduce chemicals and pesticide residues. For it to go the lengths of commissioning – and publishing the distorted results of – such a bogus survey demonstrates that, for the first time, the industry feels itself vulnerable to the kind of consumer rejection it has suffered in the EU and elsewhere.[51]

The arrogance and sheer bullying evident in the case of Quist and Chapela, and in the determination not to accept even minimal labelling of GM foods, extends beyond research scientists and consumers to the farmers who produce America's food. The fact that many of the trumpeted advantages of GM crops turn out to be either illusory, short-lived, or, at best, smaller than promised, coupled with an increasing unease about the negative reaction of overseas consumers and governments, has led an increasing number of US farmers to question their decision to go over to GM varieties. Unfortunately for them, they are discovering that it is much easier to get into GM than it is to get out of it. In one case, for example, a farm family in Indiana were warned that unless they sowed their entire farm with Monsanto's herbicide-resistant soya, they could be sued. The farmers had planted a quarter of their land with the variety, so that they could compare it with the crops they had been growing before to see whether it was worth changing over. Despite precise records of what was planted where, Monsanto refused to accept that the fact that 'their' soya was growing on much more than a quarter of the land was a result of contamination. As in the case of Percy Schmeiser in Canada – the details of which I discuss in the next chapter – these farmers were liable to be forced to pay compensation to a company which had contaminated their land.[52]

## PHARMING

The environmental problems attendant upon pharming – the genetic engineering of plants and animals to produce substances useful as pharmaceuticals – were important in breaking the wall of silence behind which biotech corporations had been able to effect their takeover of a massive slice of US agriculture. In 2002, it came to light that maize, soya, rice, tobacco and other crop plants genetically engineered to produce pharmaceuticals were being grown at 300 locations around the country.[53]

The first documented case of contamination of a food crop by genes from GMOs designed to produce pharmaceuticals came to light in November, 2002, when the USDA ordered the destruction of the whole of a soya crop, said to be valued at $2.7 million. A company called ProdiGene had been authorised to carry out tests of crops modified to produce pharmaceutical and industrial products at almost a hundred locations in the US. A vital operation following such trials is the complete removal of any residue of the crop. In the case of a batch of maize, ProdiGene had failed to do this. This may have been less to do with any negligence than a result of the fact that such a thorough removal is extremely difficult to achieve, a problem which dogs the producers of GMOs. Some seed remained, contaminating the following year's crop, which happened to be soya. The contamination was discovered after the soya had been harvested and was in storage in a grain elevator. The exact protein involved in the contamination was not revealed, but ProdiGene's research has involved an Aids vaccine, a blood-clotting agent, a digestive enzyme which, according to Friends of the Earth, 'can be used in leather-tanning or to produce insulin', an enzyme used as an industrial adhesive, an experimental oral vaccine for hepatitis B and another aimed at preventing gastroenteritis in pigs.[54]

Even before this incident, a task force established by the Biotechnology Industry Organisation (BIO) had recommended a voluntary embargo on pharming in food-growing areas, a ban also demanded by retailers' groups. The industry had responded by temporarily embargoing the use of food and feed crops genetically engineered to produce pharmaceuticals in three major food-producing states and parts of six others, from 2003. ProdiGene was ordered to destroy the crop and fined $250,000, while BIO adopted its task force's recommendations, calling for biopharmaceutical crops not to be planted in the corn belt unless contamination through outcrossing was, due to the nature of the species' reproductive system, impossible. ProdiGene, meanwhile, posted a bond of $1 million against any further contamination and agreed to reimburse the USDA for the costs. Company Chief Executive Anthony Laos was quoted as saying that his firm had 'learned some valuable lessons' and that they 'expect[ed] the enhanced compliance program' under development in 'close cooperation with USDA to set the benchmark for regulating the entire industry'. Food distribution companies were, in the meantime, calling for a ban on the use of food crops for pharmaceutical production through genetic engineering. Under what was described as 'intense political pressure', however – meaning the vested, megabuck-backed interest of a small group of corn-belt politicians – such demands were quietly forgotten, while BIO weakened its stance, now stating that it would merely 'encourage and invite alternative approaches... that would deliver at least equivalent assurances for the integrity of the food supply and export markets'. This slide from unequivocal if reluctant commitment in the face of initial panic, to corporate blather is a common feature of the industry's style.[55]

Unfortunately for those who favour the 'self-regulatory' (for which read 'no regulation') approach, pharming of plants to produce drugs and vaccines has provoked far more alarm than any other aspect of agricultural biotechnology. Whilst there may be health concerns regarding the genetic engineering of food plants, it is easier to persuade the public that the addition of, say, a gene from one vegetable to another in order to improve its flavour or give it a longer shelf life is safe than it is to convince them that it is safe to manipulate a plant so that it produces a medicine which, if it is anything like the drugs with which people are familiar, may be life-threatening rather than life-saving if taken inappropriately or unwittingly. In the wake of ProdiGene it became known that crops were being grown in the open environment and that the regulatory system and the scientific knowledge on which it is supposedly based were inadequate to guarantee safety. It was not a question of whether contamination would happen but when and where. USDA records show that by November, 2002, the company had received 85 test permits for experimental open-environment trials of such crops. Some of the permits covered more than one site, so that in total 96 field trials were ongoing or about to go ahead.[56]

Reporting plans to cultivate GM maize engineered 'to produce proteins and enzymes for use in the production of insulin and other pharmaceuticals' in Colorado, the *Denver Post* noted that 'the anxiety of farmers across the state has spread like ragweed pollen on a stiff wind'. Farmers were concerned not only for their own health but for their livelihoods. 'Our export customers as well as food processors like Kraft and others have said that they have zero tolerance for the drug corn. One kernel in an entire shipment will disrupt the export supply and cripple the industry,' one arable farmer was quoted as saying. Another ridiculed the idea that the GM corn could be contained: 'Corn is very promiscuous. Gene drift will happen. I guarantee it.' Rules requiring that pharmed crops be grown at least half a mile from plants cultivated for food were described as 'laughable...You have to wonder... if anybody has told the USDA about the birds and the bees.' Add to this, 'the likelihood that farm machinery, farmers, dogs and assorted wildlife would have contact with both crops, and the notion of containing the genetic material on one corner of fertile Mother earth is absurd'.[57]

In 2001 corn not approved for human consumption was detected in taco shells marketed under the brand name StarLink. This was something of a scandal, but the industry was able to claim that no actual health problems resulted, though how they could be confident that this was so is unclear. The modified corn which had found its way into the taco shells was approved for animal feed, but not for food.[58] In any case, with the aid of a compliant media, the storm blew over, but people tend to remember it whenever biotech companies publish proposals to produce pharmaceuticals by tweaking the genes of plants grown in the open. Clearly, if what had been detected in StarLink had been something with powerful pharmaceutical properties, public reaction would have been very different. As the *New Scientist* commented, 'It would take just one case of contamination in which a person was harmed by a crop laced with a potent drug for the entire nascent industry to face shutdown.'[59]

Pharmaceuticals are, almost by definition, based on molecules the ingestion of which has powerful effects on the body. Ingested at the wrong time, or unknowingly, or by vulnerable people, they can – obviously, uncontroversially, scientifically, if you like – cause problems ranging from mild discomfort to death. Whilst it is the profit motive which drives biotech corporations and those in their pay to ignore or deny such facts, other sections of the food industry, equally concerned for their profits, cannot afford to do so. The Grocery Manufacturers of America (GMA), for example, responded to the ProdiGene crisis by asking the USDA to ensure that plants which are normally used for food or feed are not used for pharming, while the president of the National Food Processors' Association described it as 'alarming' that 'at the earliest stage of the development of crops for plant-made pharmaceuticals, the most basic preventive measures were not faithfully observed', an oversight which 'very nearly placed the integrity of the food supply in jeopardy'.[60]

In response, the USDA did tighten its regulations on pharming of plants. Stricter separation of pharmed from conventional plants, in terms of both distance and planting time, separation of machinery used on pharmed and conventional crops, a sevenfold increase in the number of inspections required for each pharmed crop, all formed part of the package of reforms. This did not, however, satisfy the GMA, who continued to demand that only crops which are never, in their conventional form, used for food, be pharmed, and that companies involved in pharming be required to disclose the details of the genetic modifications involved.[61]

In the wake of what even pro-biotech publications referred to as a 'fiasco', no one was arguing that existing rules were adequate. ProdiGene spokespeople argued for a case-by-case approach, taking into account whether a substance produced by a GM plant occurred naturally and had been consumed by human beings without ill effects. Trypsin and aprotonin, for example, an enzyme and an enzyme inhibitor normally obtained from cows, have never been seen to create a problem, though ProdiGene has stated a willingness to conduct further studies to demonstrate their harmlessness. If they were able to do so, they argue, then it would be safe to grow plants producing these substances near to food plants, as contamination would, if it occurred, cause no health problems.[62]

Monsanto, however, and many others involved in pharming, have demonstrated a willingness to go beyond the USDA's clearly inadequate regulations, employing sterile male plants that produce no pollen, satellite monitoring of fields, strict segregation of farm machinery and processing equipment, and a range of other safeguards. In one instance, Monsanto ensures that no food plants are grown for two years in any field which has previously been used for pharming. In what Monsanto spokesman Jon McIntyre describes as 'a closed-loop system completely outside of the commercial grain system', instead of food plants, they cultivate a variety of cotton resistant to a herbicide which kills any leftover maize.[63]

Not everyone is impressed by this apparently socially and environmentally friendly attitude. Many who stop short of a total rejection of pharming argue that it would be mere common sense to ensure that pharmaceuticals are produced from modified plants which are never used as food. Others are unconvinced that the technology is worth the risk. As Jane Rissler of the Union of Concerned Scientists points out, 'they've yet to prove they can grow these crops safely anywhere, in any country'. Others, including some who are themselves involved in genetic engineering, are cynical about Monsanto's motives. Charles Arntzen, a scientist at Arizona State University who has developed an oral vaccine against a common cause of diarrhoea using engineered tomatoes and other plants, uses sealed greenhouses with fine-mesh screens and other features designed to ensure that, as Arntzen puts it, 'No insects or seeds can get in or out unless we let them.' Yet he believes that Monsanto's high-tech approach to safety is designed mainly to kill off the competition. 'It raises the entry barrier to their competitors,' he argues. 'Looking at satellite data every day and having all that dedicated equipment – a little company can't do that.'[64]

By 2000 the image of the agricultural biotechnology industry had taken so many blows that Monsanto felt it necessary to begin a major propaganda campaign. The industry's complaint that hostility to it was based on a lack of understanding of science was not, however, reflected in the content of the prime-time TV ads which shunned explanation of such matters in favour of a mixture of schmaltz and hype much more familiar to American viewers. Monsanto is merely helping farmers to protect the crops which feed the nation and doctors to treat its diseases. There was absolutely nothing whatsoever to worry about.[65]

## PATENTS

The laxness of US regulation of biotechnology and its products is both reflected in and bolstered by a patent system which differs markedly from that of European countries, and in ways which critically affect the industry. The first and major difference is that in the United States there is no scope for refusing a patent on the grounds that it is unethical, no equivalent of the European Patent Convention's provision that excludes anything deemed to be contrary to public order or morality. Nor is there any specific provision excluding plant and animal varieties, or anything produced by a biological process. As we have seen, such provisions have been partly circumvented in the EU by some creative interpretations of the law. In the US, however, there was far less to circumvent. It must be remembered that the granting of a patent does not imply approval, however, or even permission to go ahead with the manufacture of the object in question, or with placing it on the market. An object may be patented and then promptly banned. It is merely that the patenting system has never, in the US, been a favoured instrument for the interpretation or enforcement of morality. Nevertheless, the laxness of the patent system and its lack of concern with ethics has undoubtedly made life easier for the US biotech industry. The oncomouse, for example – a GM mouse programmed to develop cancer – could not be patented in Europe because it was a new variety of animal; in the US, this exclusion did not exist.[66]

### Plant Patent Act

Since 1930, the Plant Patent Act had allowed breeders to take out patents on new varieties, provided these could be reproduced asexually and were not propagated by tuber (thus excluding potatoes). However, the exclusion of sexually reproducing plants (on the grounds that their progeny would not be identical, and therefore could not be covered by any traditionally defined patent) made the measure of limited value to professional breeders. Fewer than 1,000 patents were issued in the first 20 years after its passage.[67]

In 1970 Congress extended the possibility of intellectual property rights in plants, but, as in Europe, this was achieved outside the patent system and offered a more limited form of protection. Technological progress meant that one of the obstacles to such protection, whether through patents or not, was being overcome. Only by the early 1970s and the development of DNA-based identification techniques had it become possible to identify a particular plant as one of a patented or otherwise protected variety by anything other than superficial means.[68]

### 'Patents on life'

The non-patentability of life was being called into serious question. In 1973, a biochemist in the employ of the General Electric Company (GEC) filed for a patent on a bacterium modified to destroy oil slicks. The claim was rejected, but on the grounds that the bacteria were insufficiently different to those found in nature, rather than on any general principle. On appeal, the Patent Office accepted the argument that they were indeed sufficiently different, but rejected the claim on the broader grounds that, outside the limited terms provided by the Plant Patent Act, living organisms were not patentable. By this time a similar claim had been filed by scientists working for the pharmaceuticals firm Upjohn. They had developed a modified fungus which generated an antibiotic more efficiently than did the parent organism. This claim had also been rejected. Appeals to the highest patent authority, the United States Court of Customs and Patent Appeals, were registered in both cases. The Upjohn case was heard first, in October, 1977. They won, though two of five judges voted against them, a pattern of voting which was repeated the following March when the GEC case was heard. The patent on life had not quite arrived, however. For technical reasons, the two cases had to be heard again, though the only result of this was that the majority in favour of accepting the applications increased to four to one. Following the original ruling, lawyers at the Patent Office had decided to appeal to the Supreme Court. They were motivated less by a hostility to patents on living organisms than by a recognition of the possible negative commercial consequences of continuing uncertainty. The Supreme Court was the only body able to give a ruling which could not be overthrown by a higher court, and thus the only court able to end the potentially destructive uncertainty. In fact, Upjohn effectively withdrew their patent application, so that the Supreme Court heard only the case involving GEC and its modified bacterium. In 1980, the Supreme Court ruled in favour of Chakrabarty, the GEC scientist who had taken out the patent.[69]

Uncertainties remained. Did the ruling apply only to micro-organisms, or did it cover plants and animals? Experiments with transgenic animals were already proceeding apace, and the possibility of producing substances with commercial potential from such animals was clear. In 1984 the US Patent Office (USPO) issued rulings in the case of a modified mouse and an oyster. The mouse had been engineered to contain a cancer-provoking 'oncogene', though the patent covered 'any transgenic mammal, excluding human beings, containing in all its cells an activated oncogene introduced into it – or an ancestor – at an embryonic stage'.[70]

The Patent Office rejected both claims: the oyster on the mundane ground that 'the innovation was obvious to anyone schooled in the art of oyster breeding'. The oncogene technique, however, was rejected on the more controversial grounds that the precedent case had not established the patentability of higher animals. A further application, for a patent on a GM maize seed, was also refused. All three were appealed, however, and in each case the Patent Office's own Patent and Trademark Appeals Board upheld the patentability of life. The maize had been turned down on the grounds that Congress had clearly intended plants to be protected under the systems established by the Plant Patent Act and the Plant Variety Protection Act (PVPA); the Board ruled that this was not the case. The oyster lost its appeal, but in ruling against it the Board made it clear that they did not wish to call into question the patentability of living organisms, other than human beings. This clearly pointed to success for the oncogene, which in April 1998 duly came.[71]

Numerous objections were raised to these developments. The House of Representatives sub-committee which dealt with patents held hearings on the issue, raising concerns about animal welfare, the economic implications of patentability, the adequacy of the exclusion from patentability of human beings, and the usurpation of the rights of Congress to determine the scope of the patenting system. In fact, virtually all of the concern over patents on life focused on animals, making it above all an ethical debate. Patents which effectively extend the 'rights' of biotechnology companies to the descendants of any patented seed sold have since become commonplace, and did not become controversial until the overbearing behaviour of

Monsanto and others revealed how potent a weapon they provide to big business in its age-old war to screw as much out of the farmer as it can. A moratorium on animal patents, however, has been proposed on numerous occasions, including in the wake of the Beltsville scandal, when a pig was genetically engineered to grow bigger, faster. The experiment was so spectacularly 'successful' that the unfortunate animal was unable to lift itself from the ground and died after a short life of unbroken suffering.[72] Congressman Benjamin Cardin pointed out that, by the criteria accepted, a two-headed animal (one cat, one dog) would qualify, due to its utility as a circus freak. Small farmers would suffer, warned a witness from the Wisconsin Farmers' Union. Animal patents would 'shift the profit motive for livestock improvements from the family farmers, who have used the classical breeding practices over the years, to the giant corporations which have the resources to use...DNA research for their own benefit.' Such voices, however, were drowned out by the bland reassurances offered by agribusiness and the biotech industry. Their arguments, 16 years on, are wearily familiar: biotech would feed the world, eradicate disease and make US farmers and research scientists rich. It was safe, and far from increasing animal suffering, it would reduce it. Small farmers would benefit from the public disclosure that patenting required, and from the ability to 'produce leaner beef...at lower cost'.[73] A bill exempting farmers from any requirement to treat the offspring of patented animals as themselves patent-protected passed in the House but was not taken up by the Senate, and therefore failed. When reintroduced in 1990 it was defeated in the House. Since then, Congress has rarely returned to the issue, and only around 25 patents on animals have been granted, most of them on laboratory rodents.[74]

In 1991 Craig Venter, who would go on to fame and fortune with the Human Genome Project, proposed the patenting of human gene fragments. Venter's laboratory had sequenced fragments of DNA known as 'expressed sequence tags' (ESTs). These ESTs could be used to identify the genes of which they were part. His application for a patent failed, though in turning it down the USPO did not explicitly rule out all patents on human genes. This provoked some alarm amongst those who had resisted patents on living organisms, and led by the anti-biotech campaigner Jeremy Rifkin, they demanded that Congress act to outlaw patenting of human beings or any part of them. Congress, however, the majority of whose members had now been convinced, by whatever means, of the economic potential of biotechnology, was not inclined to listen. Despite assembling a broad coalition which managed to unite religious leaders and feminist activists, this movement was unable to withstand the powerful wind blowing in the opposite direction. By the end of the century, the question of the patentability of life was settled. The Patents Office saw no role for ethics in arriving at its decisions. If anyone were to decide that a particular process or product were unethical, it should be the elected representatives of the people. Congress had every right to ban anything it liked, provided such restrictions did not conflict with the Constitution, a question which did not here arise. Whether something were patented or not was irrelevant.[75]

In 1995, encouraged by President Clinton, the Patents and Trademarks Office (PTO) further relaxed patent requirements for biotech-based drugs. Until then, biotechnology companies could establish the necessary criterion of 'usefulness' only on the basis of clinical trials on the target species – so that a pharmaceutical aimed at human beings would have to undergo full trials, at great expense, before a patent could be applied for. The *New Scientist* commented at the time that 'Under the old rules, biotechnologists were caught in an impossible position. Before they could apply for a patent, they had to conduct clinical trials. But until the drug was protected by a patent, investors were reluctant to put up the money for trials.' Why this position was more 'impossible' than that of anyone attempting to raise cash to cover the expense of research, development and testing is unclear, for biotechnology is surely not the only field which requires a lot of front-loaded cash before any returns can be expected. Moreover, if the government's assurance that human tests would continue to be required before a drug can be marketed was to be trusted, then it is hard to see what investors had gained. The move was, nevertheless, aimed at giving a boost to an industry which had 'hit rock bottom' in 1994, an 'investment drought' having 'left many companies seriously short of cash'.[76]

## REPRODUCTIVE AND THERAPEUTIC CLONING

The curious mixture of authoritarianism and permissiveness which characterises many areas of American law can also be seen in the US authorities' attitude to cloning. There are no federal laws, for instance, against reproductive or therapeutic cloning, only a ban on federal funding of any research using any stem cells which do not come from a number of authorised lines. Although this law is irksome, not least because it appears that less than half of the embryonic stem cell lines authorised for use are actually viable, it has not led to a tailing off in research. This may be partly because the federal authorities have interpreted the ban on federal funding somewhat loosely. If a researcher works mainly from federal funding, he or she may work on new, unapproved ESC lines or even create new ones, provided the lines themselves have not been developed with federal money. This is helpful, but as a political fudge rather than a scientific or ethical decision, it may not last. Any tightening of the rules would bring massive problems, as many of the available stem cell lines made use of mouse stem cells to promote growth, which makes the FDA, suddenly awake to the dangers of zoonoses, reluctant to authorise their use in patients. The majority, moreover, have been described as 'poorly characterised, hard to obtain or unsuitable for certain applications'. In addition, technological progress means that stem cells of greatly improved quality can now be produced. Whilst labouring under these restrictions is frustrating for researchers, they can at least console themselves that so far attempts to pass more comprehensive legislation have failed, with the vicious anti-abortion lobby sinking its fangs into the issue, their 'pro-choice' opponents divided, celebrities weighing in on both sides of the argument, most sane Americans being able to see strengths and weaknesses on all sides and no side able to carry enough votes in Congress to win the day. The uncertainty has, however, had a measurable impact on research in the field and the investment required to fund it. As Sir Paul Nurse asked, 'What sort of signal does it send out when the private sector can do anything and the public sector is restricted? How can you take such legislation seriously?' In one case, the Juvenile Diabetes Research Foundation announced plans to spend $20 million on research using ESCs. The funds would go, however, to institutions in Britain, Sweden, Australia and Singapore, 'where the regulatory climate and public opinion are more favorable than in the US'.[77]

#### Gene therapy

The possibility of developing useful gene therapies is hampered by these restrictions on the use of embryonic stem cells. In addition, those involved in experimental gene therapies have been seen, on more than one occasion, to do things which appear to put their research interests ahead of those of the patients involved, in other words to commit the very act of instrumentalising another human being which opponents warn and many others fear will be inherent in this approach.

The most famous example of this was Jesse Gelsinger. Eighteen at the time of his death, he had been born with the rare liver disorder ornithine transcarbamylase (OTC) deficiency. He was, however, able to lead a fulfilling life with the aid of drugs. OTC deficiency means that the liver cannot handle ammonia, and sufferers are vulnerable to brain damage, and can fall into coma. The disease can be life-threatening, but Jesse Gelsinger's life was under no threat from it. What would kill him was not even an attempted cure, but an experiment to see whether a gene therapy approach could lead to one. The poignancy of Gelsinger's death was increased by the knowledge that he had volunteered for the treatment knowing that he would not benefit directly. His participation in the trials was an act of pure philanthropy, and one of the many distressing aspects of the affair is that it may discourage others from such selflessness.

The gene therapy to which Jesse was subjected involved the use of a viral vector, a virus adapted to carry genetic material of use to the patient. In this instance, the virus of choice was an adenovirus, the bug that gives you the common cold. The doctor in charge of the trials, James Wilson, favoured the adenovirus because it is able to infect almost every cell in the body, and ought therefore to be able to deliver the replacement gene to where it was needed. The point of the trial was not to find a cure for OTC deficiency *per se,* but rather to test Wilson's theory that the adenovirus would provide an effective gene therapy vector. Success, however, could lead to a cure not only for OTC deficiency but for hundreds of diseases.

The trial participants were given varying doses of the adenovirus and Jesse Gelsinger was given the biggest. He was warned to expect some cold- or flu-like symptoms. Instead, on the second day after the viruses were injected into his body Jesse went into a coma from which he never woke. The adenovirus had provoked a massive immune reaction in his body which led to organ failure. Precisely why this happened is still not known. However, once again it pointed to the fact that there is much about the behaviour of organisms at the cellular and genetic level which is poorly understood. Although the adenovirus had been injected directly into the circulation of the liver, it found its way into other organs, a completely unexpected occurrence. In addition to such unpredicted complications, however, Wilson's team had failed to follow standard rules for clinical trials, allowing Gelsinger to enter the trial despite elevated levels of blood ammonia, and failing to notify the FDA about side-effects seen in patients. Whilst such failures are not specific to gene therapy, they seemed to many to provide further evidence of the arrogance with which so many scientists seem to become infected when they begin tinkering with genes.[78]

This disturbing conclusion was further reinforced when, in the wake of Jesse Gelsinger's death, investigators began to look at the broader picture. What emerged was that hundreds of patients had become ill or died during gene therapy trials and that all that really distinguished the Gelsinger case was that, for various reasons, it came to national and international attention. In almost every case of an 'adverse event' during seven years of trials, although researchers had obeyed the letter of the law by informing the FDA of these, this did not amount to the 'public disclosure' required by the other federal authority, the National Institutes of Health, whose Office of Biotechnology Activities monitors trials. Neither the FDA, whose reviews are confidential, nor the researchers themselves complied with this requirement. In those seven years, only 39 of 691 'adverse events' during trials using the adenovirus vector were reported to the NIH. Of course, most people entering clinical trials are ill, often dangerously so, and it cannot be assumed that their deaths are due to the experimental treatments they are undergoing. However, the FDA's statement that only Gelsinger's death can definitely be attributed to gene therapy, whilst literally true, is also clearly misleading.[79]

Jesse Gelsinger was a young man who was in most respects healthy, and by all accounts charming and bright, with the probability of a personally fulfilling and socially useful future in front of him, a future which will now never happen. His unnecessary death was the result of carelessness and arrogance, and the outrage which this provoked extended far beyond those who had previously had any negative feelings about the kind of therapy which the experiment which killed him was designed to develop. Gene therapy has saved lives as well as costing them,[80] and there are more rational grounds for scepticism than those provided by the Gelsinger case, for people die every day in hospitals as a result of mistakes, negligence and a host of avoidable factors. Yet responsible researchers no doubt feared that it would make it more difficult for them to find volunteers for trials, and not just those which involved gene therapy. Jesse and his parents had, after all, been assured that the therapy would involve no risk to his life. Under such circumstances, the authorities were obliged to act, and if they were to resist calls for a moratorium, they had better make sure that changes were more than cosmetic. Procedures were tightened and rules enforced, and so far there has been no repeat of the tragedy. The fact that this sad business led to reform is certainly to be welcomed, but whether the attitudes which led to the original failure have changed is open to doubt. Despite the reforms, the tracks which were being desperately covered in the wake of the scandal hardly brought to mind the road to Damascus.[81]

## BIOETHICISTS

Much of what passes for public discussion of the issues involved in medical biotechnology has been hijacked by 'bioethicists' in the pay of the industry. If you work for a TV or radio company and you're asked to plan a news item, discussion programme or documentary on biotechnology in health care, or if you're a journalist planning a feature or looking for a comment, then it would be natural to turn to bioethicists employed by universities to look into such issues. This is, of course, why they have almost all been bought.

Non-profit research centres and bioethics departments run by the leading universities in the field have accepted funding from industry sources, as has the American Medical Association's own ethics institute. Some publish details, others do not. Recently published guidelines are, according to dissident bioethicist Carl Elliott, 'likely to make matters even worse'. As he goes on to say, 'Given the fact that the [then] presidents of the ASBH [American Society for Bioethics and Humanities] and ASLME [American Society for Law, Medicine and Ethics] were working for Geron and DNA Sciences respectively, it is not all that surprising that the task force report endorses, without qualification, for-profit bioethics consultation to industry.' The report, in fact, approved the whole of the biotech industry's agenda, including using bioethics in advertisements, and the statement that the authors 'felt no need' to comply with the conflict-of-interest policy normally operated by the Hastings Center, the non-profit body which had commissioned it, including the disclosure of corporations or other bodies worked for and monies received. They do however, reveal that of the ten bioethicists charged with writing the report, 'eight...have performed the kinds of ethically controversial corporate consultation the report addresses'. Elliott goes on to cite other routine violations of what would seem elementary ethics by people supposedly hired to teach and research the subject at the highest level. He asserts:

> Until recently, students studying the ethics of stem cell research would not have suspected that their teacher was a consultant for Greron; scholars criticising industry-sponsored clinical trials would not have imagined that the editor evaluating their manuscript was working for Eli Lilly; and newspaper readers would not have thought that the ethicist commenting on genetic engineering was drawing a pay cheque from Celera.

The industry cannot simply rely on misleading students and the public to get everything it wants; for that, a more direct approach is also necessary. Dubbed 'influence peddling' by Carl Elliott and, when it refers to politicians and civil servants, 'regulatory capture' by others, this involves direct financial inducements, as well as more subtle bribes – flattery is often favoured – to influence the decisions of people such as 'directors of major centres, editors of bioethics journals, members of national policy committees, presidents of bioethics associations and authors of standard bioethics texts'. Through such means 'bureaucrats become tools for the industry they are supposed to be regulating, because they are dependent on industry representatives for funding, career advancement or professional respect'. Indeed, bioethicists themselves are often in precisely this position, and not only in the private sector but in the increasingly dependent universities and research centres and the increasingly infiltrated public authorities.[82]

## REGULATORY CAPTURE

This phenomenon affects far more than the limited circle of bioethicists. The conclusion of one study was that 'Flawed regulatory oversight resulted in licensing, then withdrawal, of many dangerous drugs...Established protections for human subjects in medical research...are being undermined.' One reason was that

> nine out of ten doctors on committees that develop clinical guidelines had financial ties to the industry whose products they recommend. Six of ten doctors had financial ties to companies whose drugs were considered in the guidelines they wrote. Pharmaceutical companies paid for the development of 25 percent of the guidelines.

Citing the case of breast cancer, Meryl Nass, a physician concerned about these developments, said that because 'there exists a multi-million dollar establishment that deals with breast cancer in a fairly monolitihic way, one is limited as to what questions are allowed to be asked. You can ask, but who will fund your research?'[83]

These problems are not peculiar to biotechnology. The lack of effective regulatory procedures and the collapse of the ethical imperative to separate business interests from one's professional judgement as a doctor or scientist has, however, both fuelled and fed the growth of the profit-motivated biotech industry's increasing hegemony over biotechnological research.

## FURTHER READING

John Stauber and Sheldon Rampton *Trust Us, We're Experts: How industry manipulates science and gambles with your future* (New York: Tarcher/Putnam, 2000)

Martin Teitel and Kimberly A. Wilson *Changing the Nature of Nature: What you need to know about genetically engineered food* (London: Vision Paperbacks, 2000)

Daniel J. Kevles *A History of Patenting Life in the United States with Comparative Attention to Europe and Canada* (European Group on Ethics in Science and New Technologies to the European Commission/Office for Official Publications of the European Communities, 2002)

Jeffrey M. Smith *Seeds of Deception: Exposing industry and government lies about the safety of the genetically engineered foods you're eating* (Fairfield, IA: Yes Books, 2003)

Bernard D. MacGaghey and Thomas P. Redick, *Liability and Labeling of Genetically Modified Organisms* (Report of a Round Table organised by Missouri Botanical Garden) <www.cast-science.org/0002abab.htm>

# 3
# Other Developed Countries

America's claims that GMOs are sweeping the world are belied by the statistics. Outside the US and Canada, almost every country is taking a cautious approach and only one, Argentina, has fully committed its agriculture to a biotechnological future. Aside from these three countries and the EU, only Australia, China, Cuba, Egypt, Japan, South Africa, Romania, Bulgaria and New Zealand have committed significant amounts of land and resources to developing GMOs. As for medical biotechnology, although the international consensus favours a cautious approach, including a ban on human reproductive cloning, legislative frameworks have developed more slowly than the science towards which they are addressed.

## CANADA

Canada's approach to biotechnology largely reflects that of the United States, though on the international stage it has shown itself more willing to negotiate and compromise. As in many other countries, restrictions on medical and reproductive biotechnology are relatively stringent, though Canada does allow the use of supernumerary embryos.

The Canadian pharmaceutical industry is a significant player and was, until quite recently, able to take advantage of domestic privileges, in exchange for which it was able to supply Canadians with cheaper, generic drugs. The fact that this system has been brought to an end has a part to play in the story of Canadian biotech, and I will return to this subject below. The country's vast agricultural sector has, however, generated by far the most controversy of any of biotechnology's various applications, and most of that controversy has revolved around GMOs: who should be allowed to grow and not to grow them, and whether consumers had the right not to be forced to eat them.

### The strange case of Percy Schmeiser's canola

In 1997 Percy Schmeiser, a Saskatchewan farmer, was annoyed to discover that his fields had been contaminated by the genetically engineered canola (known in Europe as rape or rapeseed) increasingly favoured by his neighbours. Schmeiser was particularly irritated because one reason why he had shunned the GM varieties was the fact that he had devoted a great deal of time and energy to improving his canola through sophisticated cross-breeding methods. Contamination threatened to undo years of hard work and ingenuity.

When inspectors from Monsanto, the corporation which had sold his neighbours their seed, discovered that their Roundup Ready variety was growing on the land of a farmer who was not contracted to them, they accused him of patent violation and suggested that he had obtained the seed fraudulently and for gain. Even though charges to that effect were dropped, a Canadian court would later find Percy Schmeiser guilty of the lesser charge of having Monsanto's product on his land and failing to advise the company of its unauthorised presence. Schmeiser was found guilty because, according to the presiding judge, the source of the Roundup Ready canola was 'really not significant for the resolution of the issue of infringement'.[1] Nor was it important that Schmeiser had neither sold the resultant seed nor sprayed the canola with Roundup, and therefore had clearly sought to derive no economic benefit from the unauthorised plants. All that mattered was that a variety of seed whose patent belonged to another party was growing on his land, and that he had taken no steps to alert the patent owner. He was guilty, and had to pay a large fine, as well as damages.

Wherever fields of it are grown commercially, canola's pretty yellow flowers festoon roadsides and the edges of woodland. Its small seeds are blown on the wind and carried on the boots of men and women and on the bodies of animals. They blow from trucks during transport, and stick to the tyres of farm vehicles. In one Scottish study, GM contamination was identified 26 kilometres from where the parent plants were cultivated.[2] Seeds can remain dormant for up to ten years, germinating at any time of year sufficiently warm to allow growth. Moreover, though the species is predominantly self-pollinating, outcrossing – pollination of one individual by another – also occurs, and pollen can be carried many kilometres by insects. Through all of these means, a high rate of genetic contamination can be expected.[3] In addition, once your land has been contaminated, or if you once deliberately cultivate a herbicide-tolerant GM variety, you will find it extremely difficult to get rid of it. Its seed will persist in the soil, emerging next growing season as 'volunteers'. In parts of the world – such as western Canada – where GM canola is widely cultivated, contamination is now so pervasive that seed companies will not guarantee their products to be 100 per cent GM free. As we saw in Chapter 1 on the European Union's legal framework for GMOs, the EU no longer requires that a product be absolutely free of GM material before it may be legally sold without a label to the effect that it contains such. Monsanto, one of the loudest voices insisting that 100 per cent free cannot be guaranteed through current methods of analysis, is nevertheless prepared to prosecute farmers who fail to achieve it – unless those farmers inform them that Monsanto-owned genes have ruined their crops.[4]

### The GM takeover of Canada's food

As the major player in biotechnology outside the United States and European Union, Canada provides an interesting case study, almost a microcosm of what has happened as a result of the rise of biotechnology over the last quarter of a century. Changes in the regulatory system, in funding, in the relationship between public and private sectors, and between government, universities and industry, as well as the reform of the patent system and developments in the broader economic and political environment, all reflect global shifts in power, ideology and relations of production and trade.

Just as is the case in the US, food based on genetically modified organisms has stealthily taken over Canada's supermarket shelves without the knowledge, still less the permission, of the vast majority of its people. Internationally, Canada is the third largest producer of GM crops in the world, and yet in common with the US requires no labelling of GM products.[5]

The Canadian Food Inspection Agency is responsible for approval of both field trials and marketing of food and of animal feed, using powers derived from the Seeds Act, the Plant Protection Act, the Feeds Act, the Fertilizer Act and the Health of Animals Act. In some cases it shares responsibility with Environment Canada (the federal environment ministry) for the application of relevant sections of the Canadian Environmental Protection Act, and with Health Canada under the Pest Control Products Act and the Food and Drugs Act. Its brief is to assess the potential impact of the release or marketing of GMOs or GMO-based food and feed, taking into account biodiversity, human, animal and plant health, and the degree of risk of gene flow or other sources of contamination. By 2001, 51 crops had been approved for placing on the market, contributing to a situation which sees, by the government's own estimation, GMOs present in 60 per cent of processed foods.[6]

### The development of the biotech industry in Canada

According to Devlin Kuyek's thorough study of the rise of biotechnology in Canada, *The Real Board of Directors,* this development had its roots in the early 1980s, when the federal government, concerned that Canada might be left behind in an impending biotech revolution, commissioned a so-called Task Force on Biotechnology which led, in 1983, to the adoption of the National Biotechnology Strategy (NBS). This occurred in the broader context of the election of a right-wing government committed to moving national policy in a clear neoliberal direction based on the Reaganite-Thatcherite idea that the 'role of government' is 'to deregulate industry, enforce intellectual property rights...and subsidize high-technology research and development to attract foreign investment'. Biotech was seen as a key sector, and its representatives moved 'deep into Ottawa's corridors of power'. The virtual destruction of Canada's bizarrely named Progressive Conservatives following a decade in government made little or no difference, as the Liberal governments which followed continued and deepened this commitment.[7]

From the start, public money was invested heavily in private corporations, though universities also played a crucial role both in research and development and in what Kuyek describes as 'mobilising government support for biotech'.[8] Initially the private sector showed little interest. By the early 1980s, US biotech was attracting considerable private investment, but Canadian researchers continued to have to rely almost exclusively on public money. When private sector interest did arrive, it was in the form of MNCs, eager, as Kuyek puts it, 'to scoop up any research with commercial potential.' The justification, and the counter-argument to anyone churlish enough to wonder why the taxpayer was getting the bills while the private sector shareholder picked up any profits, was an early example of the 'partnership' rhetoric which has since become the wearily familiar leitmotif of neoliberal plunder of the public purse. 'Partnership', in this particular dialect of nuspeak, meant an acceptance of the dubious, and certainly unproven, proposition that, as Canadian Minister of Science William Winegard put it, 'R&D is more effectively carried out by the business and university communities where it is industry-led and results-driven.'[9] That the private sector must be allowed to define just what 'results' might be desirable was assumed, and the obvious truth that these would be results quantifiable in Canadian dollars politely left unspoken.

Like its agricultural counterpart, the Canadian biotech pharmaceutical sector developed in the context of the NBS. In the early stages, it was not seen as a priority, as Canada was committed to the defence and development of its own generic pharmaceuticals industry which clearly would have no use for biotechnology until, at least, the cycle of research, development and patenting had been played out elsewhere. Canada supplied affordable medicines to its people on the basis of a system of compulsory licensing which was not abandoned until 1983, when the new Progressive Conservative government saw fit to put the interests of the international pharmaceutical industry – meaning, for the most part, US 'big pharma' MNCs – before those of the country's people. Using public funds, two Canadian corporations, Connaught and Allelix, gave the country its own indigenous industry.

Reliance on public funds, and the fact that these could be transformed, effectively, into risk-free seed capital leading to profitable products, necessitated a huge lobbying effort and gave birth to the alliance of right-wing (including the 'New Labour' variety) politicians, big pharma, agribusiness and biotech which has since developed into the extraordinary international lie machine which generates biotech industry propaganda. The results were palpable: 1987 saw both a measure to strengthen patent protection and attract greater foreign investment to the sector, and the establishment of the National Science and Technology Policy which listed biotechnology as one of only three sectors regarded as 'paramount to Canada's international competitive position'.[10] In 1988 the government followed this up by nominating a number of centres of advanced biotechnology research in different cities across the country, with Montreal spearheading the pharmaceutical sector, Saskatoon chosen for its role in the development of canola as a major commercial crop, and Vancouver taking the lead in health care and forestry.[11]

Canadian biotech was nevertheless slow to get going, employing only just over seven thousand people in the entire country a whole decade after the establishment of the NBS with its promise of heaps of genetically enhanced jam tomorrow. The author of a survey conducted in 1995 quipped that 'Several respondents noted that it was premature to use the word profitability in relation to Canadian biotechnology firms since none had as yet achieved that enviable position.'[12] Although the sector was now attracting more private investment, continued input of public funds, the tailoring of many areas of policy to suit the needs of the biotech industry, and the intertwining of private and public interests was what kept the industry going and even, albeit at an uninspiring pace, growing.

By the end of the 1980s, however, much of the research being conducted in Canada was either under contract to multinational corporations or conducted directly by firms which had now been taken over by those same MNCs. Aided by friendly government policies which improved patent protection for branded pharmaceuticals, made large amounts of public money and publicly employed expertise available to every branch of the biotech industry, deregulated key market sectors and embraced the principle of substantial equivalence as an acceptable measure of safety, Canadian biotechnology – and the activities of US and other foreign biotech companies in Canada – began to flourish.[13]

The decision, taken on the basis of the bogus 'substantial equivalence' principle, that no new regulatory agencies were needed to oversee biotechnology or its products, neatly reproduced developments south of the border. Decisions regarding the safety of GM foods would be the responsibility of Health Canada, which must conduct a safety assessment of any food which has not previously been marketed in Canada, or which is produced by means of a process not previously used in Canada. Under the Canadian Food and Drug Regulation, however, the concept of 'substantial equivalence' is recognised, being assessed by a comparison of the 'molecular, compositional, toxicological and nutritional data for the modified organism to those of its traditional counterpart'. This renders the system ineffective as a means of protecting the health of the consumer or the broader environment.[14]

The ministry did establish a new Biotechnology Management Team, but this was not a regulatory body. Instead its brief was to define Agriculture Canada's – the federal ministry of agriculture – '"corporate position" and co-ordinat[e] activities relevant to biotechnology'.[15] The government's position on agricultural biotechnology was thus nothing if not clear. As a statement from Agriculture Canada's Food Production and Inspection Branch asserted, 'If Canada's agri-food industry is to be competitive in global markets in the future, we must establish ourselves as leaders in biotechnology. There is great potential for biotechnology to improve the competitiveness of agriculture products through added value.'[16]

The collapse of the Progressive Conservative government in 1993 – the party was left with two seats in the federal parliament – signified no long-term change of policy in relation to biotechnology, simply, along with some shifts in emphasis, a deepening of the commitment which had been growing since the early 1980s. Although, as part of an overall budget-balancing exercise, federal funds to biotech were cut substantially, these cuts were temporary, effectively restored as part of the Liberal government's newly established and well-funded Canadian Foundation for Innovation (CFI) in 1998. The CFI also represented a centralisation of decision-making, and, according to Kuyek, it meant that 'the interests of the biotechnology industry in Canada became more dependent than ever on support within the uppermost circles of federal power'.[17]

This process is similar in some ways to what has occurred in the European Union, with Ottawa, like Brussels, gaining power at the expense of more accessible provincial/national capitals and, at the same time, unelected bureaucrats and 'advisers' from industry taking over decision-making roles once reserved for elected assemblies and the ministers and civil servants directly answerable to them. Under the new arrangements, a body such as the Canadian Institute for Advanced Research (CIAR) which 'brings the elite of university scientists and the business establishment together on several programs that [it] believes will be key to future economic development)' would gradually acquire more influence than government ministers.[18]

The following year's budget pumped large sums into what was defined as 'innovation' in science and technology, much of it going to biotech, as well as introducing a generous 'Scientific Research and Experimental Development' tax break, worth, by the government's own estimate, $1.4 billion per year. Industry spokespeople were overjoyed, one noting that 'Canadian R&D incentives are now clearly, clearly, the most favourable in the G7.'[19] Public funding now plays an important role in relation to most currently available applications of biotechnology, including the important forestry industry.[20]

Identified in 1997 by Prime Minister Jean Chrétien[21] as a key sector, biotechnology policy would henceforth be presided over by no fewer than seven ministers whose task would be to make Canada a world leader in the field. Money from the federal coffers poured into the industry through the ministries responsible for agriculture and industry as well as the CFI.[22]

Three years after Chrétien's establishment of his team of seven, the biotech industry's millennial celebrations were given extra fizz by the announcement, in a speech by Finance Minister Martin, of the establishment of Genome Canada, which was to receive an average of over $100 million a year in its first three years of operation. According to its own website, Genome Canada's 'overriding objective' is 'to coordinate genomics research to enable Canada to become a world leader in a few selected sectors that are of importance to this country. Such as health, environment, forestry and fisheries.'[23] At the heart of Genome Canada's brief stands what is clearly a conflict of interest, as the same body is charged with the tasks of promoting genomics through co-ordination of the different public and private bodies involved, providing financial support and developing research infrastructure, and helping to attract private sector investment; and, on the other hand, 'Ensur(ing) leadership in ethical, environmental, legal and social issues related to genomics.' How it is supposed, in practice, to perform this latter function is unclear, though who will do so – an advisory committee consisting entirely of people with an interest in the industry's wellbeing, and entirely lacking in representatives of, for example, consumers or environmentalist groups – is not.[24] Such conflicts of interest are to be found throughout the regulatory bodies which are supposed to oversee biotechnology in Canada and elsewhere. In forestry, for example, the Canadian Forestry Service 'both promotes GM research and checks for risks'.[25] Once again, we see the mentality of 'if it's good for business, it's good for everyone' undermining not only democracy but, in some cases, practices which predate anything which could be defined as democracy but which were once seen as elementary aspects of sound administrative practice.

Under these arrangements power is not only accumulated in Ottawa at the expense of the Provinces, it is also transferred from the federal parliament to government, and thence into the hands of the Prime Minister. Shadowing the team of Cabinet Ministers established by the Prime Minister to oversee biotechnology policy is a further team of Deputy Ministers, the Biotechnology Deputy Ministers Coordinating Committee (BDMCC) whose public utterances demonstrate that they have been chosen entirely for their enthusiasm for 'Life Sciences'.[26]

Biotech's developing 'partnership' with government – or, if you prefer, the federal government's prostration to its interests – would be much more difficult were sustained and informed resistance to develop on the basis of deep and broad public concern. In the one case where such resistance is evident, for example, the federal government has refused to go along with 'patents on life', rejecting Myriad's application for a patent of the oncogene successfully patented in the US, a decision upheld by the Supreme Court in 2002.[27] Keeping such resistance in check requires different skills from those which manipulate the governmental and bureaucratic decision-making processes directly, and these in turn demand, as in the US and EU, different kinds of lobbying organised through different bodies.

Established in the 1980s as part of the NBS, the National Biotechnology Advisory Council (NBAC) had been the principal conduit of opinion and influence between industry and government, but had, until the late 1990s, little or no direct contact with the public. To meet what it now perceived as a pressing need to generate public support, the government, in Kuyek's words 'morphed NBAC into the Canadian Biotechnology Advisory Committee [CBAC]', expanding its membership, reducing the proportion of direct industry representatives, and placing it under the aegis of Industry Canada, with a responsibility, however, to report to the BMCC. The CBAC's role is therefore twofold: to advise the government on how to handle 'sensitive issues'; and, by engaging directly with the public, 'to provide the semblance of consultation'.[28]

Again, this exercise in sham participation will be familiar to anyone who has followed developments in the EU. It is part of the style of modern parliamentary democracies, with their ubiquitous jargon of 'stakeholders' and their endless forums, workshops and 'national debates'. Backed up by numerous semi-official bodies, described by Kuyek as 'hybrid lobby groups at the fringes and outside of government', the CBAC serves to draw potential dissenters into the establishment, disguises the undemocratic nature of the real decision-making process, and helps the government to avoid writing unworkable law. One vital lesson learnt by neoliberal politicians from the eventual demise of Thatcher – probably the most dictatorially minded leader of any ostensible democracy in modern times – was that consultation of some kind is necessary if bad law – not 'bad' in the moral sense, but simply in terms of efficiency, workability, and the achievement of goals – is to be avoided. As Kuyek explains,

> It is understood from the outset that the government and the advisory bodies share a common agenda. The advisory bodies, and the government itself, are only there to act out the roles of and make a few adjustments to a script that, in many ways, has already been decided upon behind closed doors.[29]

Nothing could better describe, for example, the role of the European Commission's unconvincing exercises in 'consultation' with 'stakeholders'.

If, as I suggested at the beginning of this chapter, Canada provides us with a handy microcosm of what biotechnology has done to the world since its first stirrings as a commercial phenomenon in the late 1970s, then the major changes which it has brought about can in turn each be exemplified by a single instance of reform. The development of patent law can be fully understood through an examination of Canada's Bill C-22, which moved the country from a system based on compulsory licensing[30] to one which clearly prioritised the interests of pharmaceutical multinationals. The fact that the federal government in 1997 closed or sold off almost all of its health research laboratories provides a stark instance of the way in which, internationally, research funding and the control which goes with it has moved from public to private, though it goes without saying that much of the actual money continues to come from the wage packets of taxpayers. The difference is that, firstly, the tax take from the industry itself has been reduced, allowing the private sector to keep more of the profits it generates and therefore to determine directly how much of those profits are reinvested in research and just what kinds of research are carried out; and, secondly, where public funds are handed over, the industry itself has increased its ability to say who should get them and how they should be spent. This in turn forms part of a broader change, also evident internationally, in regulatory systems, changes which have made the industry all but self-regulating, removing more and more powers from elected institutions, or ones which are at least answerable to such, and placing it in the hands of bureaucrats and appointees, many of whom turn out to have direct links to the industry which they are supposed to be supervising. All of this has been aided by the general context in which it has taken place, one of globalisation, deregulation, liberalisation and privatisation. The retreat of the state, and thus the retreat of any semblance of democracy or possibility of popular control, from whole areas of economic life is nowhere more evident than in relation to those areas of the economy of most relevance to biotech. The consequence is that resources are allocated less and less on the basis of popular need and increasingly on the basis of what is termed 'competitiveness'; that they are directed towards certain economic activities or areas of industry privileged for the size of their political clout rather than in the interests of the 'nation', still less of its people, least of all of those people or activities most in need of these resources.

The overall problem, in Canada as in the United States, the United Kingdom and increasingly elsewhere, is that in the world of biotechnology public and private sectors have become so intertwined that their interests have come to appear identical. Canadian biotechnology shows just where the spurious logic of 'if it's good for business, it's good for everyone' leads – specifically, in this case, to 20 years of propping up an industry which could not exist in a genuine 'free market', which produces things which are at best unwanted and at worst dangerous. Environmental writer Stephen Leahy spoke for many Canadians when he complained that

> Federal and provincial governments have long had a love affair with genetics, pumping billions into the biotech biz since the early 1980s... So, 20 years later and how many breakthrough products has biotech produced?...The industry consistently overhypes the benefits and downplays the risks of a revolutionary new technology.[31]

This penetration of government, and of the public decision-making process as a whole, by the biotechnology industries also has its effects in the broader regulatory environment, hindering, for example, attempts to extend to consumers the right – to some extent won in Europe, but wholly absent from United States' law – to reject food containing or produced from GMOs should they wish to do so.

**Weakness of regulation**

As consumer and environmentalist groups began to demand labels, and the system of field-to-fork segregation and traceability needed to make them credible, the industry trotted out its usual arguments: segregation was impossible, consumers are ill-informed, and campaigners are anti-science weirdoes who use irrational arguments to generate hysteria; GMO-based foods are 'substantially equivalent' and thus the expense and difficulty of labelling and traceability is not worth the candle. A bill which would have required GM foods to be labelled was narrowly defeated in October 2001. As more scientists and mainstream voices joined the chorus of demands for labelling, however, the government was forced to react. Sensibly, rather than deny that, in a perfect world, labels might well be a perfectly reasonable thing to require, Ottawa shifted ground. An Agriculture Canada statement issued in November 2002 stated openly that the reason why labelling had been rejected was because of fear of upsetting the United States. US hostility to labels meant that labels were out, simply because the US was the biggest customer for Canadian food exports and would not allow them into the country were they to carry labels along the lines of those now adopted in the EU. Moreover, as the biggest importer of food into Canada the US would never agree to label its GM-based foodstuffs. 'The adoption of [a] mandatory labelling system by Canada could have a significant impact on its trade relationship with its largest agricultural trading partner', the statement declared, while a 'disjointed approach with the US on voluntary versus mandatory labelling could place both trade and investment at risk'. Some trade officials went further than this, pointing out that 'a mandatory labelling regime in Canada would be challenged by the US as a new trade barrier that contravenes NAFTA [North American Free Trade Agreement] rules'. The only real difference with the US was that the Canadians supported voluntary labelling, which the US has opposed with almost as much vigour as it has a compulsory system. Under the sponsorship of the Canadian Council of Grocery Distributors, in cooperation with the Canadian General Standards Board, a system of voluntary labelling is under development.[32]

## AUSTRALIA

Australia first established a voluntary system of controls over genetically modified organisms as early as 1975, long before the first commercial applications. Under this system, a non-statutory advisory body, the Genetic Manipulation Advisory Committee (GMAC), assesses GMOs and their production, deciding whether they pose a risk to human, animal or plant health or the environment. This system of risk assessment was not, however, backed by legislation, so that its major purpose was to alert biotech firms themselves to dangers which might lead them into breaches of general criminal or civil law.

### The Gene Technology Act

In 1999, the federal government went some way towards recognising the weaknesses inherent in this approach, setting up the Interim Office of the Gene Technology Regulator (IOGTR: later the Office of the Gene Technology Regulator (OGTR)) under the aegis of the Department of Health and Aged Care. The OGTR's brief is to work with State and Territory Governments, other federal government agencies, the private sector and non-governmental organisations (NGOs) to develop and implement a new national regulatory system for GMOs.

The following year saw the introduction of the Gene Technology Act with the stated objective of protecting human health and the environment by identifying and managing risks posed by the application of modern biotechnological methods to agriculture. The Act established three key advisory groups whose job is to work with the Gene Technology Regulator. A Gene Technology Technical Advisory Committee (GTTAC) replaced GMAC; GTTAC's scientific advice is supplemented by further advice from the Gene Technology Ethics Committee; and by the Gene Technology Consultative Community Committee, which advises on public opinion and includes representatives of consumer groups, environmentalists and researchers. The work of the advisory committees and the supervision of the regulatory system is presided over by a Ministerial Council representing the Commonwealth (federal) government and those of the States and Territories.[33]

The Gene Technology Act covers every aspect of the process of creating, cultivating and marketing GMOs and GMO-based products, from health and safety within the laboratory to environmental releases and marketing, though some individual products are covered by sectoral legislation. Before a GMO can be released, a risk assessment must be conducted similar to that demanded in the EU and Canada.[34]

Like that of the EU, Australia's approach is based explicitly on the precautionary principle. Unlike the USA and Canada, and in common with the EU, Australia has recognised the need for an entirely new regulatory regime. This is true for food safety regulations as well as for laws governing deliberate release. Here, Australia and New Zealand share a common agency, Food Standards Australia New Zealand (FSANZ). In 1999, FSANZ adopted new rules under its Food Standards Code, rules specific to GMOs and GMO-derived foodstuffs. The rules require a pre-marketing safety assessment but are based on the principle of substantial equivalence. In relevant cases, the National Registration Authority for Agricultural and Veterinary Chemicals (NRA) and the Therapeutic Goods Administration (TGA) are also involved in co-ordinating marketing approval of GM products.[35]

Australia has also established a system of transparency designed to make available to the public information on GMOs and GM products, including details of applications, and conditions placed on the granting of a license for deliberate release or marketing, reasons why such an application has been refused, location of GMO release sites and details of the GMOs themselves and the methods used to produce them. This information is stored and made available to the public through what is known as the 'Record of GMO dealings and GM products', described as 'a complete list of all dealings with live, genetically modified organisms...approved by the Gene Technology Regulator...and of all GM products approved by other Regulators'.[36]

Regulation is backed by a system of 'monitoring and compliance activities', including flowering and harvest inspections and unannounced 'spot checks'. Members of the public are encouraged to report suspected breaches of the regulations, and these are followed up with inspections. The results of the monitoring and compliance inspections are published by the OGTR in a quarterly record.[37]

In principle, no one may have any 'dealing' with a GMO – the peculiarly Australian general term covering research, cultivation, manufacture, production, commercial release and import – which is not licensed by the Regulator, either for contained use or for deliberate release. However, exemptions for contained use may be granted under what is known as the Notifiable Low Risk Dealing (NLRD) system, where past experience shows this to pose minimal risk. In all cases the 'dealing' must be included in the register of GMOs. The major difference between NLRD and the US system is that it cannot be applied in the case of deliberate release.[38]

There remains an effective moratorium on commercial growth of GMOs in Australia. The bulk of the Regulator's work is therefore concerned with research projects conducted by universities, public sector bodies and small firms. Since 2000, when Australia officially adopted a 'national biotechnology strategy' and established a Biotechnology Innovation Fund with an initial stake of $20 million, public investment has been aimed at transforming laboratory studies into commercially viable products.[39]

These policies have been under tremendous pressure since their introduction, with the federal government under attack by state governments eager to grab a slice of what they are convinced is a juicy biotech pie. On the other hand, public suspicion of agricultural biotechnology in particular is increasingly spilling over into visible anti-GM activism.[40] Opponents have followed the lead of broad movements in New Zealand and Switzerland, calling both inside and outside parliament for a five-year moratorium on all releases, whether experimental or commercial, as well as on marketing and imports, to allow the safety of the technology and its products to be assessed.[41]

The Gene Technology Act has itself been subjected to severe criticism by environmentalists and consumer advocates who argue that it is inadequate to the task of protecting Australians, their agriculture and broader environment from the dangers posed by GM foods, GM crops and contamination from GM varieties. Particular criticisms focused – as they would a couple of years later in Europe – on the lack of an effective liability regime backed up by compulsory insurance. The Green Party leader, Senator Bob Brown, for example, put the case during the passage of the bill for an alternative approach based on a five year moratorium 'to apply to the import of all GM products and the release of all GMOs'. In addition to arguments based on environmental and public health considerations, Brown pointed to the possible commercial disadvantage should Australian farmers lose their ability to guarantee a GMO-free product. In addition to a moratorium, the Greens sought to diffuse decision-making, moving it closer to affected communities by giving local authorities as well as State and Territorial governments the right 'to prohibit the release of GMOs within their jurisdictions'.[42]

#### Medical biotechnology

Legislators have generally taken a wait-and-see approach to medical biotechnology. In September 2002, however, Australia's House of Representatives voted to ban all human cloning, including therapeutic cloning. Shortly after this, they were presented with a bill allowing stem cells to be harvested from spare embryos, as is the case in many other countries. Australia was a leader in the field of stem cell research and those involved in it understandably ran an intense propaganda campaign in the run up to the vote. Other scientists, however, opposed the campaign, arguing in the case of one submission to parliament that it was 'scientifically premature' to work on human ESCs until the techniques involved had been properly tested on animals. Supporters of their use were discredited when one leading researcher presented a video to parliament that turned out to have been falsified. Instead of human ESCs, the rat, which regained the use of its hind legs, turned out to have been injected with foetal cells. The exposure of the ruse jeopardised A$46 million which had been provisionally earmarked for stem cell research and the establishment of a new National Stem Cell Centre.[43]

In the end Australia adopted a law allowing supernumerary stem cell use, the centre got the go-ahead, and the scientist who had presented the misleading video was even made its head. Australian stem cell research could continue, albeit under greater restriction than in the UK, where researchers may create embryos for research purposes.[44]

## NEW ZEALAND

New Zealand's regulation of biotechnology has followed a similar course to that of Australia. In May 2003 its government adopted a new Biotechnology Strategy, which Science Minister Pete Hodgson described as being 'about developing the biotechnology sector with care'. The idea was to build on the work of the Biotechnology Taskforce, part of a general 'Growth and Innovation Framework', announced in February 2002 under which biotechnology was identified as one of three major sources of potential growth. Research activity is primarily, though not exclusively, in the public sector, involving universities and state bodies known as Crown Research Institutes. In the government's view, 'New Zealand has a strong base in biotechnology that stems from its unique access to and overlap with biomedical and primary sector research', bringing together findings from a number of areas and sectors to generate innovative ideas and products. An independent analysis commissioned by the government listed 'nine important sectors which offer great potential for further development' including, unsurprisingly, agriculture and pharmaceuticals, but also the especially controversial areas of transgenic animals and 'nutraceuticals'.[45]

### The Hazardous Substances and New Organisms Act (HSNO)

New Zealand has in recent times seen rising opposition to GMOs, opposition so strong that government has been unable to ignore it. This was not, however, initially the case. Before 1994, New Zealand had conducted a number of experimental releases of GMOs. None had required special permission, and none had provoked controversy. It had eventually been recognised that some form of regulation was imperative, however, and in that year the government introduced the Hazardous Substances and New Organisms (HSNO) Bill. After a tortuous path through the legislative process which took four years, the bill became an Act. Its most important provision was a requirement that prior approval be gained for all research involving genetic modification. The body with the right to give or withhold such approval was the Environmental Risk Management Authority (ERMA).[46]

A number of features favoured the development of genetic engineering in New Zealand. Firstly, the country had a right-wing government fully committed to a neoliberal 'if it's good for business it's good for New Zealand' credo. Secondly, and partly as a consequence of this, state-controlled scientific research, largely administered by a number of Crown Research Institutes (CRIs), was hugely influenced by a corporate agenda which sought the development of profitable products. The idea that public funds might best be spent on projects for which private industry, which needs things to sell, would understandably ignore, was regarded as hopelessly out of touch.

Debate over the HSNO Act, first proposed in 1994 but not finally adopted until almost three years later – drew public attention to the potential problems associated with GMOs. At the same time, a series of related scandals woke New Zealanders up to the fact that something strange was growing at the bottom of their gardens – or, in one case at least, of their beaches.

The first scandal resulted from the discovery that imported GMOs had been in NZ supermarkets for some time. They carried no distinguishing label, and the government appeared to have no plans to require them to do so. This revelation led to the further discovery that a total of 238 deliberate releases of GMOs, some of them animals, had been allowed to occur and that regulation of field trials was almost non-existent. By the time that the Greens, a significant minority party in New Zealand, made the issue a central plank of their 1999 general election platform, there remained no possibility that it could be swept back under the carpet.[47]

The governing coalition, made up of the right-wing National Party and the equally right-wing Maori-based party New Zealand First, at first refused to countenance demands for a labelling scheme, though exceptions might be made in the case of foods which were 'substantially modified'. In the face of growing demands, however – and confronted by the usual difficulty of explaining what, if the foods themselves were fine, was so problematic about labelling them – the government changed tack, agreeing to introduce labels for, in principle, all GMOs and GM-derived foods. The announcement provoked the usual bullying response from the US, with Ambassador Josiah Beeman warning that they would take a 'dim view' of any such requirement.

The unannounced GM crops and unlabelled GM foods were themselves the object of a 'dim view' from the New Zealand public, but it was the case of the genetically modified fish which really swung public opinion firmly behind the campaign for, at the very least, tighter regulation and more transparency. King Salmon had been breeding genetically modified salmon off the coast of South Island for four years, having voluntarily sought government approval and also received some financial backing from the state. The location, moreover, was a beauty spot popular with tourists, fishermen and lovers of water sports.

In April 1999, King Salmon's experiment came to public attention when the Green Party leaked a document which showed that the firm had employed a public relations company to help it deal with the fact that some of its fish had developed clear deformities, their heads growing unusually large and lumpy. It turned out that the PR people whom they had employed had also been hired by the CRI's Gene Technology Information Trust (GTIT), and that its brief had not been, as publicly stated, to inform the public impartially about biotechnology, but to persuade it that biotechnology was a good thing. The firm, known as Communication Trumps, was employed to run a programme called Gene Pool, which involved a 'road show', an information pack, and all the familiar paraphernalia of modern information (and brainwashing) campaigns. Almost all of GTIT's near $200,000 funding went on Gene Pool. Further investigation then uncovered the fact that most of this money came from companies with a major financial interest in the future of biotech in New Zealand. Many were semi-public or recently privatised bodies such as the NZ Beef and Lamb Marketing Bureau and NZ Kiwifruit, but $27,500 had come from Monsanto. GTIT was closed down a month before these details were finally revealed, after a great deal of stalling, when in October 1999 its former head appeared before a parliamentary select committee. The details of the funding were what was regarded as scandalous, but private sector agenda-setting and attendant conflicts of interest had been made inevitable by the requirement that CRIs conduct 'user-pays research' rather than pursuing a programme based on perceived public benefit.[48]

### The Royal Commission on Genetic Modification

In 2000, the government established a Royal Commission on Genetic Modification (RCGM) to report on options. The Commission, headed by former chief justice Sir Thomas Eichelbaum, also included a biochemist, a Maori health expert and a representative of the church. During the year that it was working – from spring 2000 – the Minister for Environment and Biosecurity, Marion Hobbs, called a moratorium on applications for new releases of GMOs, backing up her request for a 'voluntary' pause with a threat to impose a compulsory version if it was not met. She got her way, and although experimental releases were reinstated, a moratorium on commercial cultivation remained in place until October 2003. The Commission then proceeded to hear a wide range of views, a feat accomplished in the face of intermittent bullying of expert witnesses from pro-biotech groups when such experts did not share their views.[49]

When, a year later, the RCGM delivered its report, it contained something to please all sides, inevitably meaning, of course, that it was guaranteed to upset everyone. Setting out its approach, the Commission noted that 'while most [New Zealanders] were comfortable with genetic modification for medical purposes, many strongly opposed other uses'. In the Commission's view, this is because 'First-generation genetically modified crops have shown few obvious benefits for consumers.' Respondents to their consultation exercises 'stressed that the safety and certainty of the science have yet to be proved' and that 'world consumer preferences are against use of genetic modification in food'. Because of this, 'New Zealand should keep its options open…[and] proceed carefully, minimising and managing risks.' Commitment to research and development of conventional and organic farming methods should not be neglected. These conclusions are, in fact, almost identical to the position of the European Commission, which holds that 'different production systems should not be seen as being in opposition to each other, but rather as contributing in their own ways to the overall benefit of New Zealand'. Ethical difficulties are complicated in New Zealand by official recognition of the need to respect traditional Maori values, which casts the EU's practical conviction that there is a single group of 'European values' which are consensual and virtually synonymous with truth and reason in an interesting light.[50]

The Royal Commission recommended a strong system of regulation for GMOs, moving the country's laws firmly into line behind the emerging international consensus, and away from a flirtation with the United States' broad interpretation of 'substantial equivalence'. The major device for achieving this would be a series of amendments to the HSNO Act. In addition, the Royal Commission recommended that 'public research be allocated to ensure organic and other sustainable agricultural systems are adequately supported', that Maori be included in the consultation process influencing allocation of research funds, and that 'public research funding portfolios be resourced to include research on the socio-economic and ethical impacts of the release of genetically modified organisms'. In relation to crops, special measures were recommended governing the use of Bt varieties, the protection of honey from contamination, and any proposed cultivation of GM trees, as was 'a labelling regime to identify genetically modified seed, nursery stock and propagative material at point of sale'. An 'industry code of practice' should be drawn up to deal with the problem of co-existence and to protect the integrity of food and feed. The law should 'allow for specified categories of genetically modified crops to be excluded from districts where their presence would be a significant threat to an established non-genetically modified crop use'.[51] In the wake of the report, the government imposed a two-year moratorium on crop trials to allow issues of safety to be discussed and addressed.

In relation to medicine, the Royal Commission proposed 'that all gene therapy, whether in the public or the private sectors, require formal medical ethical oversight'. A new body, to be called Toi te Taiao: the Bioethics Council, should develop, *inter alia*, ethical guidelines for xenotransplantation involving genetic modification technology. Any medicines or 'pharmaco foods, functional foods or dietary supplements', whether for human or animal use, using genetic technology should be subject to strict control.[52]

Patents were also to be treated cautiously. Patents on human beings and 'the biological processes for their generation' should be ruled out, though experience from the EU and US suggests that this is, in itself, no guarantee that all human life processes would become unpatentable. The Commission also wanted to ensure that Maori were fully consulted regarding what should and should not be patentable and 'that New Zealand be proactive in pursuing cultural and intellectual property rights for indigenous peoples internationally' and, in particular, 'pursue the amendment of the World Trade Organization Agreement on Trade-Related Aspects of Intellectual Property Rights and associated conventions to include a reference to the avoidance of cultural offence as a specific ground for exclusion or reservation'.[53]

### The New Zealand Biotechnology Strategy

It is interesting to turn from these recommendations to the proposals for implementation contained in the government's *New Zealand*

*Biotechnology Strategy,* published in May 2003. The first striking difference is the tone. Whereas the Royal Commission had been charged with developing an informed consensus, and had included a wide range of opinion regarding the likely merits of various applications of biotechnology, the Strategy was different: the government had consulted widely, the Royal Commission had been a major part of that consultation, and now it was time to act.

Nevertheless, though the tone of the *Strategy* was more positive than had been the Royal Commission, most of the latter's recommendations were incorporated into the government's regulatory proposals or supporting measures. Minister for Research, Science and Technology, Pete Hodgson, in a foreword to the *Strategy*, described the biotechnology industry as 'a sector generating biological knowledge, skills and technology that can contribute to achieving our economic, social and environmental aspirations.' He also, however, spoke of the need to have 'regard for ethical and cultural concerns' and for 'robust regulation that safeguards people and the environment'.[54]

In many ways, the balance of forces in New Zealand resembles that in the UK. The pressures on the government come, on the one hand, from a vigorous, determined and growing movement against GMOs, though there seems little disquiet about other applications of biotechnology, especially those which do not concern agriculture and where no animal welfare issues are involved. On the other, an aggressive and unscrupulous industry, backed by the US, seeks, as it does elsewhere, to muddy the debate with false assertions and to pretend that the fight is between 'science' and 'Luddism'. The difference is in the degree to which the respective governments of New Zealand and the UK have dealt with this conflict. Both are sympathetic to the industry, but the former seems to have much more genuine commitment to the kind of 'way forward' favoured, for example, by the European Commission, but consistently resisted by Britain's representatives in the EU Council of Ministers. This involves applying a genuinely tight regulatory framework in the hope that this will lead to 'public acceptance' of GMOs. It is too early to say, either in the EU or NZ, whether this has worked, though the signs so far are, from the industry's point of view, far from promising.

As things stand the government's line is to advocate, and indeed implement, tight regulation while, however, never questioning the need for the industry or its products. Thus, one of the central elements of the *Strategy* is to 'Grow New Zealand's biotechnology sector to enhance economic and community benefits', while the point of 'a regulatory system that provides safeguards' – but one which also 'allows innovation' – is to facilitate the 'development and introduction of new biotechnologies'. The decision is made: biotechnology is a good thing, but it needs supervision. A Biotechnology Sector Taskforce would 'specifically focus on the commercial growth of the sector', a sector which the *Strategy* presented as a natural development of New Zealand's traditional agricultural base, coupled with its post-war emergence as a centre of scientific achievement and innovation. Tellingly, a small potted history of Maurice Wilkins, the New Zealander who shared the Nobel Prize with Crick and Watson, appears in the *Strategy*. Though somewhat irrelevant – Wilkins is an octagenarian who has done all of his science far from home – it enhances the impression that embracing biotechnology is vital to New Zealand's national pride. No problem if you have criticisms and concerns about biotech, but to oppose its development in NZ is clearly unpatriotic.[55]

In order to counter any such tendency, the *Strategy* stresses the need for 'constructive engagement between people in the community and the biotechnology sector'. Instead of demanding, as the US, often with British support, purports to do, that all criticism of biotechnology be based on 'science', New Zealand's *Strategy* expressly commits the authorities to 'Provid[ing] opportunities to consider cultural, ethical and spiritual issues', including those likely to be of concern to Maori people. Implementing the RCGM's recommendation to establish a body called 'Toi te Taiao: The Bioethics Council', the government described the Council's task as to 'play a key role in enhancing New Zealand's understanding of the cultural, ethical and spiritual aspects of biotechnology, and ensuring that biotechnology development has regard for the values held by New Zealanders'. That these values might actually exclude some crucial applications – GMOs, for example – is not taken into consideration, however. Nevertheless, talk of respect for such values seems more than hot air, as the *Strategy* includes a commitment to amend the HSNO to allow 'the consideration of cultural, ethical and spiritual matters' even to the degree that the minister could actually ban or otherwise restrict a biotechnological application purely on these bases.[56]

The assumption is that New Zealanders' 'values' and a world-beating biotech sector will prove compatible. Again, this decision has already been taken: the Biotechnology Sector Taskforce 'has set itself 10-year sector growth targets (e.g. a five-fold increase in the number of core biotechnology companies to over 200 from 40) and identified issues for attention in order to achieve these targets'. Despite the emphasis on consultation, public involvement and repeated explicit references to the importance of generating a consensus which must include the Maori peoples, the topic to be discussed has already been settled. Clearly, it is not whether NZ should have a biotech sector, but how the country can have a world-beating biotech industry, which is at issue.[57] Research will include investigation of biotechnology impacts under the heading 'Sustainable Biotechnology'; regulation 'must achieve robust safeguards for people and the environment'; the system of environmental protection must be 'comprehensive and stringent'; the stress throughout the *Strategy* is inclusive, emphasising popular participation and shunning technocracy; but, in the end, the big decision is not open to change.[58]

In relation to medical biotechnology no such final decisions are evident, though this merely reflects the uncertainty surrounding the subject internationally. Other than joining the international movement to ban human reproductive cloning, New Zealand, in common with most other countries, is clearly undecided over the best way to control the application of biotechnology to medicine and health care.[59]

Overall, New Zealand appears to have followed the EU in adopting a strict system of regulation, coupled with a very positive public view of the industry. Although the *Strategy* is more convincing in its acknowledgement of possible problems and the public's reservations than is the EU's equivalent *Life Sciences and Biotechnology – A strategy for Europe*, it embodies a similarly restricted view of the acceptable limits of the debate.[60] GMOs will not be banned. Medical biotech will be encouraged. Laws will be enacted, many on the basis of the RCGM, to 'ensure' that the risks are dealt with. ERMA will gain new powers to impose conditions on the release of GMOs, whereas previously it could only say yea or nay. This measure in many ways exemplifies the approach: adding flexibility appears to be a concession to opponents of GMOs, yet at the same time it makes it easier for ERMA to authorise release.[61] Moreover, ERMA's approval policy is far more permissive than would seem to be in keeping with the undertakings given in the *Strategy*: in 2002, for example, it approved experiments involving inserting human genes into cattle, despite receiving 850 submissions against the proposal, against seven in favour.[62] ERMA, as well as the Ministry of Agriculture and Fisheries, has also been criticised for the laxity of its surveillance. For example, when GM-contaminated maize was discovered in conventional farmers' feeds, an independent scientist called in as a consultant concluded that 'the gaps in New Zealand's ability to deal with a similar incident had not been plugged' and that 'the issue needed to be addressed urgently'. The division of responsibility between different bodies was unclear, and there was confusion over accountability and the details of legal requirements. The result was that decisions, in this case on how to clean up the contaminated fields, took too long to be arrived at and implemented.[63]

Whether or not these criticisms were instrumental in bringing about real improvements is unclear. There have, however, been some positive developments, and at least two measures do seem to represent an unequivocal victory for opponents: a clearer attribution of liability for harm caused by any activity which does not comply with the HSNO, backed up by 'a civil penalty regime for breaches of HSNO or approvals granted under the Act'; and, in the medical field, 'mandatory ethical approval for non-established procedures, such as newly developed techniques to aid fertility, and for all research'. Set against these advances are one or two clear concessions to industry: the approval process for laboratory research involving 'low-risk GMOs' will be subject to 'streamlining', as will the approval process for GMO-based pharmaceuticals. The approval, in the face of stiff opposition, of two experimental releases of GM trees in July 2003, moreover, demonstrated that the government would not always be taking a conciliatory approach, that it was determined to establish New Zealand as a centre of plant biotechnological research.[64]

Like the European Commission, the New Zealand government has performed a fine balancing act, hoping to isolate the most determined and uncompromising opponents so that the great mass of 'ordinary people', who are seen as sceptical but not fundamentally opposed will come to accept biotechnology in all, or most, of its applications. This is clear when the *Strategy*, in conclusion, states that success will be achieved if there is 'significant movement towards sector growth targets' on the basis of 'public confidence in the way the sector carries out its work' and 'public confidence that the regulatory system provides the necessary safeguards'.[65]

Whether these goals can be achieved is open to doubt. Though the arrogant tone which characterises the US biotech industry's public pronouncements is missing, the assumption is the same: GMOs (or gene therapy, or any other application you care to name) are safe, and the only real problem is that the public needs to be convinced of this. If this is not the case, and there are real dangers attached to genetic engineering, then the strategy behind the *Strategy* becomes a pure propaganda exercise. Certainly, this is how it is seen by anti-GM activists in New Zealand and their growing band of sympathisers.[66]

The reality behind such fears was revealed when, in October 2003, parliament approved the government's request not to renew the moratorium on commercial releases of GMOs. In the debate, Greens were joined by others more concerned with the possible effects on NZ exports than they were with environmental or health-based arguments. Two-thirds of the country's export earnings come from agriculture, horticulture or forestry. Yet the moratorium's end, which had been announced in January, was confirmed. An estimated 9,000 people – a major demonstration for New Zealand – marched through Wellington to protest against the lifting of the moratorium. Also exemplary was the fact that, when, in October 2003 after fierce debate and in the face of enormous public demonstrations, the government lifted the moratorium on GM crops, it followed the announcement swiftly with an assurance that no contamination at all would be acceptable in non-GM products.[67]

## SWITZERLAND

The system for regulating genetically modified foods in Switzerland in many ways resembles that which until recently prevailed in the European Union. A GMO, a food that contains GMOs, a food additive or processing aids can be marketed only if the public authorities are satisfied that, in the light of current scientific knowledge, it poses no danger to health. As with the EU, GMOs are assessed on a case-by-case basis. Anyone wishing to market a GMO must supply detailed information regarding the modification method employed and the characteristics of the resulting GMO. The process, however, revolves around the question of substantial equivalence. Approvals last five years.

Labelling is similar to the system abandoned by the EU in 2003, when it adopted the new Regulation on Labelling and Traceability. Foods must only be labelled if modified proteins or DNA are present. In other words, if differences cannot be established by verifiable testing procedures then no label is required. The 'paper chain' of documents allowing GMOs to be traced from field to fork plays no role in the Swiss system.

When in 1999 labelling regulations were reformed, a threshold of 1 per cent contamination was adopted below which foods containing GMOs would not have to be labelled as such. Unlike in the new EU regulation, Switzerland continues to allow a 'GM free label', or rather one which states that the contents of a can or package were 'produced without genetic engineering'. Such claims must be backed by documentary evidence. The 1 per cent threshold also applies to these products. One interesting feature of Swiss law is that products made from animals may only be labelled as 'produced without genetic engineering' if the animals have not been fed on GMOs, a proposal which, in a slightly different form, was specifically rejected by the European Parliament and Council of Ministers. Approvals are issued or denied by the Federal government, while laboratories administered by the Cantons are responsible for policing the system.

There have been a number of attempts to force the government to introduce a more restrictive system. Under Switzerland's unique system of direct democracy, the government is obliged to call a binding referendum to amend the constitution if more than 100,000 citizens sign a petition demanding it. A wide-ranging referendum on biotechnology – the Gene Protection Initiative – was held in 1997, but the voters rejected by a 2:1 ratio its proposals to ban all transgenic animals, all releases of GMOs, and all patents on 'genetically modified animals and plants, as well as their constituents, the procedures employed thereby and the products obtained'. Novartis and other Swiss-based companies active in biotechnology were able to convince the great majority that any such restriction would be extremely harmful to the country's economy and that this outweighed any risks to health or the environment, which were in any case hugely exaggerated.[68]

In a country which is home not only to direct democracy but to a number of major pharmaceutical corporations, a vigorous green movement and a traditionally minded small farmer class, however, the issue would not go away. The 2002 law on GMOs described above was passed in the federal parliament's Lower House by a narrow majority, and only after the original proposal had been strengthened by making rules for commercial release stricter than those covering experimental releases. As a reaction to the new law, a number of different types of organisation announced that they were voluntarily adopting a GM free policy, that they would neither cultivate nor trade in GMOs or products derived from them. These included the powerful Swiss Farmers' Organisation, which represents 90 per cent of its potential membership; the Swiss Milk Producers, which claims 100 per cent representation of dairy farmers; the Swiss Bakers' Federation; Migros and the Co-op, which between them control 70 per cent of the country's retail food industry; and a newcomer to the same industry, Carrefour. In addition, a broad coalition of environmental NGOs, the Green Party and small farmers' groups launched a campaign for a new referendum. The goals of this referendum would be narrower, its principal aim being to impose a five-year moratorium on new releases and imports. This measure, ostensibly designed to allow more time for the accumulation of evidence on the effects of GMOs on the environment and health, had previously been backed by a special government commission before being rejected by parliament.[69] The following May, however, when the question cropped up again in the form of an amendment to an agricultural appropriations bill, the Lower House changed its mind and voted in favour of the moratorium. When, a month later, the Upper House rejected the amendment, the Lower House also reversed its earlier position, killing the measure. By this time, moreover, the pro-moratorium coalition had already begun collecting signatures. By September 18, the required 100,000 had been achieved. Polls indicated that a moratorium would gain overwhelming backing were the vote to be held immediately, but the constitution gives the government five years to hold a referendum demanded by popular initiative.[70]

## JAPAN

Japan's Ministry of Health, Labour and Welfare (MHLW) is responsible for granting food safety approvals for biotech products, while the Ministry of Agriculture, Forestry and Fisheries (MAFF) deals with authorising commercial and experimental growth, and animal feed safety approvals. MHLW also tests imports to ensure that they do not contain unauthorised GMOs or products in which such GMOs have been used. A legal requirement for safety assessment of GM foods was introduced in 2001. The process, which applies to all foodstuffs and food additives produced by recombinant DNA techniques, requires an assessment of the parent organism and the effects on it of the introduced genetic material. The chemical composition of the resultant GM food or additive is compared with that of the parent organism to see whether it can be classed as 'substantially equivalent'. If not, toxicological data must be provided on the basis of which a decision for approval of any non-substantially equivalent food or additive is taken.[71]

Food labelling responsibilities are divided between the two ministries, according to product. In 2001 a labelling scheme was introduced. It is, however, so generous in its exemptions as to be virtually useless, as the threshold for GMO presence in unlabelled products is 5 per cent. The permitted 'non-GM' label is slightly less misleading, as to qualify for it a manufacturer or distributor must be able to provide documentation showing that the ingredients were demonstrably non-GM – that they come from identity-preserved (IP) sources[72] – at each step of the process leading from field to shop. Nevertheless, provided this is done, a contamination level of 5 per cent is permitted, unless a lower level would make the GM product one of the principal three ingredients. Unapproved varieties are, however, not permitted, even as trace contaminants. The only exception to this is a 1 per cent threshold for unapproved animal feed varieties which have passed a safety assessment elsewhere, provided the country issuing the assessment is approved by MAFF as having a system at least as strict as Japan's own. In contrast to EU law, products must be labelled only if they contain detectable modified proteins or DNA. The labelling requirement applies only to a list of some thirty specified foods, though the government claims that this is merely because GM products will not be found in other foods which will be added should such presence become possible. The UK Consumers' Association has criticised this approach on the grounds that 'it is partly based on a listing system largely restricting labelling to situations where novel DNA and/or protein will be present in the final product'. Giving the example of 'the inclusion of corn starch products and the exclusion of corn flakes and some potato products', it argues that the law is in practice inconsistent.[73]

The list of approved GM products and releases includes not only soya and maize, but aduki beans, broccoli, canola, cauliflower, cucumber, melon, papaya, potatoes, rice, sugar beet, tomato, a number of flowers and cotton. Apart from potatoes and sugar beet, which are only imported, these are grown as crops.[74]

The discovery at the end of 2002 of imported maize contaminated with StarLink maize (unapproved for human consumption in the United States, its country of origin, where it is, moreover, no longer grown) led Japanese importers to seek alternative sources. This proved problematic, however, as such sources were hard to find and often unreliable in terms of quality and certainty of supply, as well as uncompetitively priced. Japan's wheat buyers have, however, indicated that they will seek alternative sources should the US approve GM wheat varieties.[75]

Until 2001 Japanese academics were not allowed to establish commercial enterprises without first giving up their posts. In that year, however, the law, which went some way to protecting the integrity of Japanese academic life from the corruption which is now the norm in the United States and increasingly so elsewhere, was scrapped. This was, according to the director of Tokyo University's Human Genome Centre, Professor Yusuke Nakamura, because while until recently 'Japanese scientists have considered it inappropriate to use basic scientific information in industry', now 'there is a different wind blowing and people realise that interaction between academia and industry is important for Japan'.[76]

In 1999, the government pledged up to $650 million a year to the industry. The intention, according to the *Financial Times*, was 'to help Japan breathe new life into the sluggish genomics research base.' A number of 'Millennium Projects' were chosen, including research into the rice genome, genetic susceptibility to disease and bio-informatics, all relatively non-contentious areas, given that the knowledge gathered can be used for a variety of purposes. Increased knowledge of rice genetics, for example, would be as useful to an organic seed company as it would to a genetic engineer. The same case could be made for projects which may be of use to the pharmaceutical industry, such as the compilation of a database of single nucleotide polymorphisms (SNPs), genetic variations[77] in the Japanese population. Unfortunately, the scrapping of the law forbidding academics to establish their own companies will import into Japan precisely the same problem suffered in the west. Instead of research being used in socially beneficial ways, it will inevitably be distorted towards the most profitable applications.[78]

Japan has a relatively permissive approach to one of the most controversial aspects of medical biotechnology, with the Human Cloning Regulation Act, 2000, authorising research on human embryos in vitro, including procurement of human ESCs. The Act

> prohibits transfer of human embryo and human-animal cloned embryo made by somatic nuclear transfer, as well as human-animal chimeric embryos and human–animal hybrid embryos, to a human or animal uterus. A breach of this prohibition can be punished by up to 10 years'

> imprisonment and fine of 10 million Japanese yen (about US$93,000). So the Act is considered as a 'ban on human reproductive cloning'.

It does not, however, prohibit all human embryo research, but permits cloning of a human embryo by somatic nuclear transfer, and making human–human, human–animal or animal–human chimeric embryos, and even cloning embryos by animal somatic cell nuclear transfer to a human oocyte. Therapeutic cloning is therefore definitely not outlawed, as is trade in human embryos and oocytes.[79] As one critic puts it, this 'is strikingly contradictory to the fact that sale of human solid organs like heart or liver is legally prohibited by the Organ Transplantation Act in 1997'.[80]

## THE NEWLY INDUSTRIALISED COUNTRIES

Globally, the biotechnology sector or sectors can for the most part be divided into 'developed' or 'First World' countries, and the rest. As in many other fields, however, the so-called newly industrialised countries (NICs) of the Far East are difficult to fit into the haves/have-nots schema which is all too sadly appropriate for the rest of the world. Singapore, the People's Republic of China, South Korea and to a lesser extent other Far Eastern Pacific Rim countries have experienced rapid and (with a few well-documented hiccups) sustained economic growth which has elevated significant sections of their urban populations to 'Northern' living standards. They have done so, moreover, to a great extent on the basis of 'knowledge-based' industries which have required them to develop efficient systems of education. These conditions have made them conducive to the development of biotechnology industries over which indigenous forces – government agencies and local political and economic elites – have a degree of control.

### Singapore

Singapore, for example, at the end of 2003 established what it hopes will become a major centre of biotechnology research. 'Biopolis' is intended to house the five publicly funded biotechnology research institutes, but also hopes to attract foreign private sector operations in an attempt to maximise use of resources and create an environment conducive to intellectual cross-fertilisation. The Novartis Institute for Tropical Diseases plans to move there, and others will no doubt follow.[81]

Biopolis built on the Singapore Economic Development Board's decision in 2001 to allocate a significant sum to projects related to biotechnology, including the work of the institutes, other academic research, training and tax-incentives. This in itself was no radical break with the past. Singapore began promoting biotech in the early 1980s, with Glaxo conducting biotechnological research in the country as long ago as 1982. In 2000, the sector was declared to be the 'fourth pillar' of its economy, and $570 million was spent in establishing new research institutes.[82]

Singapore's efforts tend to be concentrated on medical biotechnology, a tendency further encouraged by the effects of the SARS crisis, which cost 33 Singaporean lives. Merck, Aventis and GlaxoSmithKline are all established in the country. ES Cell International, a smaller but significant company, exports human embryonic stem cell lines and is just the best known of a number of independent biotech firms running operations in the city-state. For those wishing to work with stem cells, Singapore has what the *New York Times* describes as 'one of the world's most liberal legal atmospheres. It allows stem cells to be taken from aborted fetuses, and human embryos to be cloned and kept for up to 14 days to produce stem cells.' However, it has banned human reproductive cloning and placed some restrictions on therapeutic cloning. The latter requires the informed consent of donors, while commercial trading in donated materials, including supernumerary embryos, is forbidden. Moreover, researchers are by law allowed, ostensibly at least, to refuse to work with human ESCs on the grounds of 'conscientious objection'.[83]

Singapore's authoritarian political system does not encourage dissent, which has, in any case, rarely anywhere in the world focused on research projects confined to the laboratory. Such criticism as has been heard in Singapore has been on the grounds of the economic advisability of committing major public investment to an industry whose actual returns have so far been disappointing. As the *New York Times* reported in 2003, although foreign firms have found Singapore in many ways conducive, 'local start-ups are struggling, and economists say the biotechnology investment is unlikely to yield jobs on the scale that electronics once did'. Moreover, investors 'are shying away from an industry in which products take at least a decade to develop. Increased competition is coming from less developed countries like China, India and Malaysia, which are building biotechnology industries of their own.'[84]

One of Singapore's major weaknesses is a paucity of home-produced scientists capable of working in the field. One attempt to tackle this has been the establishment of a major fund offering bursaries for students to pursue doctorates in biomedical sciences. This can be done in Singapore or at a foreign university, but recipients must promise to work in Singapore for up to eight years after qualifying.[85]

### Malaysia

Singapore's neighbours are attempting to follow her example. Malaysia recently unveiled its answer to Biopolis, an 800-acre 'BioValley' on which, by 2006, will be based three new institutes dedicated to agricultural biotech, pharmaceuticals and 'nutraceuticals', and genomics and molecular biology. Both countries point the way to a likely future of lightly regulated biotechnology sectors attracting foreign involvement and foreign investment. This to some extent repeats the experience of these middle-income NICs with other advanced technologies. Biotechnology, however, differs from information technology (IT) in the degree of uncertainty surrounding its future development. Even in the most optimistic of scenarios it can never replace IT as a generator of large numbers of jobs at all sorts of levels.[86]

### South Korea

In other parts of the Pacific Rim, similar developments are taking place, aided by similarly sympathetic governments and the resulting weak regulatory frameworks. South Korea began to take an interest in biotechnology in the early 1980s, when the establishment of the Korea Biotechnology Research Association was swiftly followed by the enactment of a Biotechnology Promotion Law. By 1994, South Korea was fully committed to biotechnology as a possible engine of growth, the year being declared 'The Year of Biotechnology'. Regulation was and remains minimal, directed more at ensuring Korea remains an attractive place for MNCs to invest than it was at protecting Korean consumers or the environment. The main event of the 'Year' was the inauguration of 'Biotech 2000' , a strand of HAN, 'the Highly Advanced National project'. Biotech 2000 committed a total of US$15 billion over a 14-year period, with the money being spread around public bodies (Research Institutes), the university sector (including academic 'Research Centres') and private industry. Little of this effort, however, has been directed towards agriculture. Instead, Korea has concentrated on developing 'fermentation technology, antibiotics, diagnostics, and hepatitis B vaccines' each of which has now reached the stage of commercialisation 'at the internationally competitive level'.[87] In 2004, South Korean scientists were the first to demonstrate that the two essential steps required if therapeutic cloning is to become a practicality – the production of cloned blastocysts (early embryos) and the development from these of a stable line of pluripotent stem cells – could be achieved.[88]

### The People's Republic of China

One country which appears fully committed to developing a major biotechnology sector is China. According to one enthusiast, the People's Republic of China (PRC) is seeking, as rapidly as possible, 'to create a modern, market-responsive and internationally competitive biotechnology research and development system'. The strategy is to use public funds 'to improve the innovative capacity of national biotechnology R&D, reform the current research system by providing better support to key institutions and incentive mechanisms, and promot[e] development and commercialization of biotechnology, thereby increasing investment in research'. In order to achieve this it has committed itself to generating 'a scale of investment not common in developing countries, as China positions itself to become one of the world leaders in biotechnology'.[89]

This strategy, not surprisingly, rests partly on a weak regulatory environment, though this weakness is uneven and regulation, as in most areas of life in the PRC, can be capricious. Although China became, in 2003, the first country to approve a gene therapy-based treatment after clinical trials, this seems to have been done according to norms which would have been accepted in the west and has certainly not provoked widespread condemnation.[90] As for agricultural biotech, field tests, environmental releases and commercialisation of transgenic plants were first subject to systematic regulation in November, 1993 when the State Science and Technology Commission issued the Safety Administration Regulation on Genetic Engineering. On the basis of this Regulation's requirements, the Safety Administration Implementation Regulation on Agricultural Biological Genetic Engineering entered into effect late in 1996. Earlier in the year, the Ministry of Agriculture had established the Office of Genetic Engineering Safety Administration (OGESA) to regulate field tests, environment releases and commercialisation of transgenic organisms.[91]

In 2001, the government adopted the Guidelines for Biosafety Management of Agricultural GMOs, a document supplemented the following year by specific rules on, respectively, Biosafety Evaluation Regulation for Agricultural GMOs, Import Regulation for Agricultural GMOs and Labelling Regulation for Agricultural GMOs. Despite the various legislative acts, these measures, which became effective in 2002, appear to be the first really effective legal regulations of the use of biotechnology in agriculture. Officially, at least, farms growing GMOs must now undergo regular inspections to reduce harm to the environment and GM foods are subject to risk evaluations. Imports, including species alien to China and transgenic seeds, are inspected for possible problems.[92]

Though not particularly hard-hitting, these rules have already upset the United States, as GMO importers must apply for certificates declaring that their products are harmless to humans, animals or the environment. US officials have claimed the import rules were based more on trade protectionism than on scientific concerns. When import controls were introduced, new orders of US cargoes of soybeans dwindled to almost nothing, though this seems to have been a temporary effect as confusion accompanying the new system's introduction was cleared up. Under pressure from the US, China announced temporary measures for GMO imports that required less paperwork and shorter approvals. It was not immediately clear whether the proposed law might be linked to the rules on GMO certificates. China had supported the development of biotechnology to enhance food production, pharmaceuticals and environmental conservation, the *China Daily* said, without elaborating. China's Ministry of Agriculture had drafted regulations earlier this year requiring all imported GMO products to be clearly labelled, the paper added.[93]

Foreign MNCs eager to open China to biotech's otherwise largely unmarketable products seem inadvertently to have provoked fears that foreigners may start to patent genetic material obtained from native Chinese products. This may be one impulse behind attempts to create some kind of regulatory system. The same MNCs complain that 'protectionism' is deterring investors and undermining export opportunities. At the same time, the government is quite open about its intention to draw up a 'secret list' that the *Wall Street Journal* describes as intended to 'try to keep the genetic makeup of certain goods out of foreign hands. Culled from thousands of medicinal herbs, plants and vegetables, the list will seek to isolate agricultural products deemed quintessentially Chinese, and safeguard their genetic material in a government-run repository'. The list will divide 'gene matter' into three groups: one 'that can be freely exchanged between Chinese and foreign scientists; material that can be exchanged under certain conditions; and material off-limits to the outside world'. The need for such a list was highlighted when Monsanto tried to patent a DNA sequence from a soya bean which had been grown in China for thousands of years, just as another US company has filed a patent for Basmati rice. The *Wall Street Journal* seems genuinely puzzled by China's reaction to this, but few others will share its bafflement. China is big, powerful and getting richer. It is thus no easy target for this sort of bio-piracy, though others are less equipped to deal with what has become a major aspect of the worldwide property grab that is biotechnology.[94]

Chinese agricultural biotechnology remains for the most part at the experimental stage, though cultivation is widespread and there is one major exception – Monsanto's Bt cotton. From 1986, the government has given financial support to research and development projects involving more than 100 laboratories and extensive field trials. Research has concentrated on resistance to pests and disease, salt-tolerance, drought-resistance, nutrition enrichment, quality improvement, and pharming projects such as the production of edible oral vaccines. All major projects have, however, been directed at developing pest-resistant grain, cotton and oil seeds. In 2000, the government stepped up its investment, pledging $600 million over a five-year period that ends in 2005. Commercial growth is another matter, and while the economy is in general opening up to outside investors, the government has actually introduced new curbs on investment in agricultural biotech, curbs which appear designed to prevent bio-piracy. There are a number of instances of commercial cultivation, with six licences having been issued during 1997 and 1998, though, interestingly, none has been granted since. Of the six, two are for Bt cotton, two for tomatoes, one for capsicum and one for petunias. Only one, to Monsanto for Bt cotton, has gone to either a private sector or foreign enterprise. This is, however, by far the most significant commercial growth, covering a massive 600,000 hectares. The licences together cover 35 different locations and involve over 4 million small farmers, with the area under cultivation growing from 100,000 hectares in 1998 to 1.6 million hectares in 2001. The tomatoes and capsicum have yet to be marketed and are grown on a small scale. In fact, no GM food has as yet been grown for the Chinese domestic market. Monsanto, Syngenta and others have submitted applications for commercial development of food crops including maize, soya and rice. Aside from GMOs, however, permission has been given for the marketing of a number of recombinant vaccines for animals.[95]

China is clearly hesitating about committing its agriculture lock, stock and barrel to the way of genetic modification. Growing international reluctance to embrace the technology and its products has worried the Chinese government to a point where it is showing less and less enthusiasm for the large-scale commercial cultivation of GM food crops. The problem, moreover, is not confined to export markets. Awareness of the possible problems is reportedly increasing in China itself, with several newspapers having published – and, one must generally assume, with official approval – articles on the issue. Research work, however, continues apace. In 2003 Chinese scientists were reported to be 'developing wheat and potato plants resistant to several bacterial diseases prevalent in domestic fields' while 'many other plant and even animal varieties are in the pipeline'. In 1999 China spent $112 million on crop biotech research, hardly a General Secretary's ransom but, to put it into an international context, almost ten times the expenditure of Brazil or India. The sum, moreover, was officially set to rise fivefold by 2005.[96]

Nevertheless, recent developments demonstrate that those who attempt to portray China as a no-holds-barred enthusiast for agricultural biotechnology are being rather selective with the truth. Better off or more powerful developing countries such as China, as well as those such as Singapore and South Korea which enjoy standards of living comparable to those found in most western countries, have seen biotechnology as a potential new motor of growth. Whether this will turn out to be the case, and what the environmental and social price of success might be, remain to be seen. We may disapprove, we may criticise, but countries such as China, Korea and Singapore, whatever pressures may be brought to bear upon them by foreign MNCs and governments, are to a large extent deciding for themselves that biotechnology should be part of their future.

This is not the case for most of the Third World. In relation to global biotechnology, as in so many other sectors, developing countries have great difficulty in determining their own course, in even taking into account what may be the needs, aspirations, feelings and beliefs of their own citizens. Instead, they are expected to dance to the tune of those who would use the problems of poverty, hunger and disease to undermine any possibility these countries may have to exercise their supposed right to self-determination. In the full range of political, economic and military methods the 'West' employs to maintain its ascendancy, few have been more transparent and shocking than this. Quite simply, biotechnology, and especially agricultural biotech, is being used as an instrument of western hegemony every bit as deadly as the missiles and bombs which have rained down on Yugoslavia, Afghanistan and Iraq. In this chapter, I have included some countries normally considered as part of the Third World, because these countries are, to an extent, 'doers' in the development of biotechnologies. In Chapter 4, we will see how most poorer countries have ended up in a more familiar role: not doers, but done to.

## FURTHER READING

Devlin Kuyek, *The Real Board of Directors: The construction of biotechnology policy in Canada, 1980–2002* (Sorrento, BC: The Ram's Horn, 2002) available online at <www.ramshorn.bc.ca>

Council for Biotechnology Information Centre for Safe Food *Plant Biotechnology in Canada* <www.whybiotech.com/html/pdf/plant_bt_in_canada.pdf>

Darryll Macer and Mary An Chen Ng *Changing Attitudes to Biotechnology in Japan* (2000), www.agbios.com/docroot/articles/2000253-b.pdf>

New Zealand Biotechnology Association website <www.biotech.org.nz/>

OECD *Regulatory Developments in Biotechnology in Canada* (2000) <www.oecd.org/document/57/0,2340,en_2649_34393_2370169_1_1_1_1,00.html>

OECD *Regulatory Developments in Biotechnology in Australia* (2002) <www.oecd.org/document/33/0,2340,en_2649_34393_1889569_1_1_1_1,00.html>

# 4
# Developing Countries

Opinions differ, both within and outside the so-called Third World, regarding the appropriateness of agricultural biotechnology as a potential solution to the chronic problems of hunger, malnutrition and periodic bouts of famine which dog many poor countries. These opinions do not, however, occur in a vacuum: they represent, and are formed by, economic interests, though other factors may have their influence. As opponents of agricultural biotechnology argue, 'The spread of GM technology across the developing world is not a neutral process but one shaped by powerful commercial interests and enforced by weapons of diplomacy.'[1]

## COULD BIOTECH PUT AN END TO HUNGER?

The biotech industry, on the other hand, touts itself as the world's saviour, conjuring visions of supercrops which will bring about a Golden Age in which the danger of going to bed with an empty belly is no more than one of grandma's folk tales. Anti-GM activists counter that the industry is resisting the development of appropriate systems of, for example, pest control, because these would be based on achieving long-term sustainability through reduced inputs of the very things it sells, warning that its real interest is simply, as one hostile scientist put it, in such strategies as 'draw[ing] up plans to manage pesticide resistance so as to conserve the markets for pesticides and transgenics'.[2] Throughout the developing world, small farmers' organisations and other groups of concerned citizens are resisting attempts to persuade or force their governments to accept agricultural biotechnology or, once the authorities have capitulated, to resist the spread of GMOs within their countries.

Beyond this, many – including, as will by now be obvious, this writer – believe that the industry's real agenda is to gain control of the world's food supply, and to use the economic and political power that this would give it to bring about their own Golden Age, of pliant governments, cowed populations, and eternal mega-profits. Standing between these two extremes are most developing countries' governments, some aid agencies and an array of expert opinion.

Some prefer to await developments; some accept that biotechnology *per se* may be no bad thing, but then look at who is selling it and realise that they have heard it all before. If MNCs, with their vast resources, were concerned to 'feed the world', they could clearly do it tomorrow, or the day after, using existing technologies. Why should GMOs be harnessed to a policy goal which, though clearly a human and humane imperative, has never before been made a priority by those who control so much of the world's food production? Yet others, agreeing with the anti-GM movement that agricultural biotechnology's current priorities represent a disgraceful waste of resources and ingenuity, draw the radically different conclusion that these priorities should be reordered, and that, by greater public investment, or by persuading or somehow forcing biotech companies to change their ways, agbiotech could indeed be put at the service of the world's poor and hungry masses.

This last is of importance, partly because it both underpins and reveals for what it is much of the propaganda generated by the industry, and partly because it is the approach which has been embraced by the United Nations Development Programme (UNDP), a body of considerable influence both in the Third World and amongst governmental and non-governmental aid agencies. All of this feeds, in turn, into the developing regulatory systems favoured by the largely commodity-producing countries of Asia. Africa and Latin America.

The UNDP's message is set out most clearly in its Human Development Report of 2001. This Report's controversial endorsement of biotech as a means of improving the world's food supply concluded that 'Throughout history, technology has been a powerful tool for human development and poverty reduction.' Biotech is no different, simply the latest in a series of advances capable of equipping people with 'better tools' which would make them 'more productive and prosperous'. The basis of progress must remain 'the market', but this must be tempered by an understanding that the 'market' alone would not 'diffuse the technologies needed to eradicate poverty', because poor people by definition lack the 'purchasing power' this requires. 'As a result research neglects opportunities to develop technology for poor people. Inadequate financing compounds the problem. Lack of intellectual property protection can discourage private investors.' We thus segue effortlessly from discussing the needs of the poor to promoting those of rich investors, which turn out in this case to be identical, a ubiquitous feature of this type of discourse and one which underpins the attitude to regulation of the UNDP and the elites for which it speaks. Even the statement that 'technology is created in response to market pressures' is misleading. No market pressures led to the hydrogen bomb, clearly, but even those technologies which end up on the market, and successfully so, are rarely mothered by necessity. Ten years ago neither I nor anyone else had yet realised that we 'needed' a mobile phone. Technologies often precede and create need, either through the manipulation of people's desires or by making it genuinely difficult to function as a member of a particular society unless one possesses certain products. Private transport, for example, has restructured developed countries so that it has become in many places difficult to get around without a car. Agricultural biotechnology is capable of restructuring the production of food to the degree that it may well become impossible for farmers to refuse to grow GMOs. If that happens those who control the technology may well acquire the power to eradicate starvation. They will also have the power to cause it.[3]

This is not the whole story, for if it were then the Human Development Report would read like just another piece of crass industry propaganda, which it does not. The UNDP is concerned to introduce biotech into Third World agriculture in a way which would be sustainable. This may be in some views (which, broadly, I share) impossible, but not everyone agrees, and not everyone who disagrees is an industry dupe. This has a tremendous bearing upon regulatory systems, especially in poorer countries, because the industry has (though it was largely dragged to this understanding kicking and screaming) realised that it can only hope to sell its wares by co-operating with those who, whilst accepting that biotech almost certainly carries huge potential benefits which poor countries and their farmers cannot afford to pass up, favour caution and even-handedness. The UNDP therefore recognises that developing countries 'face especially severe challenges in managing the risks' associated with agbiotech, and in the face of this the industry, or at least those of its representatives who can see beyond the next balance sheet, can only nod sagely and assure us all that they are there to help to achieve just that.[4]

Unfortunately at this point the UNDP's conclusions move from something at least resembling balance to a statement which is simply untrue, namely that while 'Consumers in countries with no food security problems tend to focus on food safety and environmental concerns', poor farmers have no time for such fripperies, concentrating instead on 'increasing food production and reducing input costs'.[5] In fact, these are of course the concerns of farmers everywhere, rich or poor, while the relationship between those who work the land and the broader environment in which they work is determined not simply by level of income but by a complex mix of factors in which tradition, culture, the degree and type of outside pressure and simply what is possible also play a role. Farmers are rarely indifferent to the environment, though, like anyone else, they may be poorly educated about its needs, especially if the main source of their 'education' is a brochure produced by a pesticide firm. India has, famously, one of the most vigorous environmental movements and it is not one which pits 'consumers' or urban intellectuals against farmers, but one in which farmers play an active and visible role. The myth that environmentalists are invariably well-educated (in formal, western terms), urban and, probably given to wearing sandals and funny hats is one deliberately propagated by those who see in nature only the opportunity to make profits, and it is disappointing to see a UN body pandering to it.

If the UNDP is correct, and biotechnology has the potential to 'feed the world', then there ought to be some evidence of its potential. As the authors of the Human Development Report acknowledge, feeding poor people has not been a priority for the industry, which must serve first of all the interests of its shareholders. Nevertheless, even if this has largely been for public relations reasons, major firms have co-operated and paid for research programmes whose goal at least appears to be to resolve problems connected to inadequate nutrition. Publicly funded scientists have also sought to develop products with real social benefits. Unfortunately, after two decades of such research, there is precious little to show.

### Golden Rice

One of the more farcical attempts by the agbiotech industry to persuade us to love it involved the now notorious case of 'Golden Rice'. According to the European Commission, which part-funded the development of this GM cereal, 'the most efficient way to find a solution' to the problem of vitamin A deficiency, a cause of widespread blindness and other health problems in poorer countries, 'is to alter the daily diet of many children in poor countries'.[6]

This is true enough. Only in the fantasies of biotech executives and researchers desperate for a public relations coup, however, does it mean that they should eat GM rice. Firstly, there are already many varieties of rice available which contain adequate vitamin A precursors, substances which generate vitamin A in the human body when digested. They are limited in distribution and not particularly popular, perhaps because they do not gleam white in a way which people have unfortunately come to value, a problem which would also afflict Golden Rice, of course. Secondly, the problem of vitamin A deficiency is a product of the Green Revolution, the last attempt to transform Third World agriculture from outside. Before the Green Revolution, polished rice was an occasional luxury rather than a staple. Because it keeps better than unpolished rice, and was long preferred in the West, the Green Revolution imported the milling technology needed to produce it. In addition, edible plants which had been gathered by women from the peripheries of farmland or from the fields where they grew as weeds were eradicated. These plants were a vital supplement to a diet which contained little other than rice. Anuradha Mittal of the Institute for Food and Development Policy gives the example of a village in West Bengal, where plants 'identified by Monsanto as weeds...to be destroyed by chemicals' were in reality 'used by the community for fodder, for medicinal purposes, and for food'. One of these was 'batua, a green leafy vegetable'. Batua is, according to Mittal, 'a great source of vitamin A [which] has been treated as a weed by the Green Revolution'.[7] Thirdly, aside from attempting to reverse these losses, many possible solutions to the problem exist, some of which are cheap and could be made readily available for much less than could any conceivable GMO-based food. Vitamin A can be administered cheaply in a pill or added to other foods which are already eaten. Educational programmes can point people towards available and affordable foods which they may be unaware contain the substance a deficiency of which is at the root of their or their children's illness. Finally, people cannot absorb needed vitamin A if they are generally undernourished. Vitamin A deficiency and the blindness which it can cause turn out, once again, to be simple products of poverty. Address the poverty and you solve the problem. There is no technical fix, no short cut which will allow the current obscene ill-distribution of wealth to continue whilst giving the guilt-provoking legion of blind children their sight. These children go blind for only one reason: they are grindingly poor, their parents are grindingly poor, their families, villages and countries are grindingly poor. The symptoms can be palliated, but why not eradicate the disease at source? Where would the money come from? Well, a start might have been made by putting the more than $100 million wasted on Golden Rice to better use.[8] More generally, money invested in biotech's limited, speculative and expensive shots at addressing the problem of hunger could be much better spent on research into technologies and systems with a track record of success, such as developing traditional means of pest control.[9]

It could be argued that in attempting to provide a solution to a problem which they and their forebears created, corporations are behaving responsibly. Yet would Golden Rice, even if it were genuinely capable of making up for the loss of vitamin-rich plants in the diets of poor people, truly represent an acceptable approach? In the late nineteenth century, most of the western working class lived in conditions comparable to those endured by workers and small farmers in developing countries today. Like them, they had to work extraordinarily and unhealthily long hours in order to feed themselves and their dependants, yet their diets were deficient in important nutrients, hunger was widespread and malnutrition a major cause of debility and disease. In western European textile areas, for example, working people (including this writer's ancestors) lived on a diet consisting almost exclusively of bread and potatoes, sugar and tea. As a result, many suffered from malnutrition and diseases such as rickets and tuberculosis were rife. A century or so later, most of their descendants (including this writer) enjoy varied diets, and if they suffer from food-related illnesses – principally obesity and those diseases of affluence known as 'eating disorders' – these are most likely to be the result of ignorance deliberately fostered by the same food industry which wants us all to eat GMOs. The solution to the problem of malnutrition was, and remains, the raising of people's standard of living to the point where they can afford a varied diet. If some Victorian genius had been able to improve potatoes so that, eaten exclusively, they produced a healthy, productive workforce, would this have been a philanthropic gesture or a cynical means of perpetuating exploitation? As a great potato-lover who nevertheless finds that they are best served with something other than more potatoes, I find it hard to be grateful to my imaginary boffin.

According to the European Commission, 'planting vitamin A rich fruits or vegetables is not always feasible'. Land reform and income redistribution would change this, but it is, of course, utopian to imagine that these are themselves feasible.[10]

In order to interest Third World governments and farmers in GMOs, new varieties will have to be bred which offer nutritional advantages. This is also good public relations in the west, provided you have a compliant press. Golden Rice has since been followed up by an attempt to engineer a high-protein potato. Unfortunately the 'protato' turned out to provoke deficiencies in iron and calcium. Perhaps that problem could be solved, but the purpose of all this is unclear. India already grows high-protein vegetable foods in the form of beans and lentils. People sometimes do not get enough of these because they are poor, bringing us back to the real reason for hunger, the real problem waiting to be solved.[11]

## THE (ATTEMPTED) RAPE OF AFRICA

In 2002 the United States government, relying on World Trade Organisation rules that it had itself dictated, attempted to use a temporary food crisis to force countries in southern Africa to accept genetically modified food. They did so because it was becoming increasingly obvious that the technology into which so many billions of dollars of both public and private money had been invested was a colossal error, a dead end. GM food was becoming unsaleable. In fact, you couldn't give it away. The big biotech firms need to sell goods that no one wants to buy, and as US food aid has in general been designed to rid America of its obscene and embarrassing surpluses, a tradition is simply being maintained.

*Genetically modified food crops are the key to ending world hunger.* Only in a world where most people rely for information on corporate-controlled television stations and newspapers could an idea so preposterous even be discussed. This is not principally because GMOs do not generally offer sustainable increased yields, though they do not. Nor is it because you are more likely to find willing investors for a project to develop slow-growing lawns than you are for one aimed at the needs of the poor, though this is also true. The major reason why the statement that begins this paragraph is absurd is because it could only possibly be true if hunger were caused by a shortage of food. This can be believed only by people who have never been hungry. Hunger is caused by poverty, by people not having the money to buy food, by the fact that food, like everything else, is ill-distributed. GMOs are to be used not to erode this lamentable state of affairs but to reinforce it. Self-sufficiency in food production, its political corollary of food sovereignty, and the development and application of appropriate, sustainable technologies for food production and distribution systems whose purpose is to feed the many rather than enrich the few, these are the clear and obvious routes to a world without hunger.

The United States has the power and resources to end world hunger and could do so in the space of a few years. It has shown, however, that for whatever reason – and a discussion of the competing explanations would require another book – it does not wish to do so. Though it does, in its own way, wage a battle for hearts and minds, its domination of the world's media means that it does not need to do this by effecting any change in its actual behaviour. There is no need for the US to behave seductively, and with the elevation of the far right to presidential office it has ceased to make any pretence of doing so. As the case of the scandal of its attempt to inflict GMOs on Africa demonstrates, the Bush Administration prefers to shun seduction in favour of rape.

This scandal did not appear from nowhere. An investigation by Friends of the Earth discovered that in the three years preceding the crisis, complaints about the inclusion of GMOs in food aid mounted up. In 2000, Ecuador received a food aid donation which included GM soya paste. The paste found its way into two food programmes, one aimed at children as young as six months and the other at nursing mothers. When the Ecuadorian authorities discovered this, they ordered the destruction of the GM material. They were able to continue the programmes, however, as Ecuadorian food – quinoa, beans and non-GM soya – was available in sufficient abundance. The GM soya had neither been requested, nor announced, nor was it in any sense necessary. The following year the US deliberately violated a Bolivian moratorium on GM imports. Their catch-all defence – the unproved statement that 'we eat the stuff in the States and we don't get sick' – was undermined when the illegal aid food was found to contain StarLink, the variety made famous by its having contaminated the food supply in the US, the UK and Denmark. StarLink is not approved for human consumption, even in the US.[12]

In 2001 and 2002, Colombia, Guatemala, Nicaragua, Uganda, India and Bosnia had all complained of the presence of undeclared GMOs in food aid. Then, in 2002, a food shortage forced southern African governments to request emergency aid. Spotting its opportunity, the United States attempted to flood the continent with GMOs, including unmilled maize, some of which would certainly end up being planted rather than eaten.[13]

The timing of this is instructive. In 2001 the European Union had adopted its newly restrictive approach to GMOs. This fuelled concerns that GM food, or food contaminated with GMOs, would become unsaleable on the world market. Grow them, whether deliberately or inadvertently, and you might find that you were producing food for which there was no market. These fears were not confined to developing countries, but in Africa, for which the EU is the biggest export market, they were particularly acute. It was this, as much as any health or environmental concern, which led to scepticism in most African countries as to the value of agricultural biotech as a means to improve their own food supplies. It was this, also, which the United States was determined to attack. Africa would have GMOs whether her people or governments wanted them or not. The alternative was starvation. At the Earth Summit in Johannesburg at the end of 2002, right-wing think tanks masquerading under green-sounding names and funded by biotechnology MNCs even paid local beggars to stage an entirely bogus demonstration demanding the right to grow and eat GM foods.[14]

Fortunately, African governments were unconvinced. While Swaziland, Lesotho and, after some hesitation, Zimbabwe, Malawi and Mozambique reluctantly accepted American food aid containing transgenic maize, Zambia stuck to its guns, with information minister Newstead Zimba explaining that the government had 'taken into consideration the scientific advice about the long-term effects of genetically modified foods and all related grains and we are rejecting it [that is, the food aid]'.[15]

Zambia was accused of denying food to its starving citizenry. Much of the coverage in the western press showed the marks of the dirty fingerprints of agbiotech's propaganda machine. This machine is apparently so powerful that it can overcome the evidence of one's own senses. The *Financial Times*, for example, carried a story whose headline claimed that 'Zambia turns away GM food aid for its starving' – accompanied by a picture of two children carrying food. The picture was captioned 'Loveness Chibali...and her sister Peggy take aid from a WFP centre in Zambia'. The food was non-GM, and these Zambian children were not starving. Yet this was no thanks to UN agencies who put pressure on the Zambian government to reverse its decision, refusing to order food aid guaranteed to be free of GMOs, even though plenty was available in the region. To do so would have been to abandon the position that GM foodstuffs were risk-free.[16]

The Organisation for Economic Co-operation and Development (OECD), however, demanded that the United States cease its pressure, accusing them of using food aid to favour the export of American farm products.[17] As we shall see in the next chapter, the Cartagena Protocol on Biosafety gives every state the right to refuse imports of 'living' GM foods, which would include unmilled maize. Although it had not come into force at the time that the Zambian government took its decision, this was in reality purely a practical matter to do with the fact that writing legislation into statute books takes time. The Protocol was signed, both by Zambia and by almost all UN members, though not, needless to say, by the US.[18]

The United States' attempted rape of Africa did not, technically, represent a breach of the Cartagena Protocol. The Protocol did not receive the 50 ratifications it needed to come into effect until the summer of 2003, but in any case the US as a non-signatory was therefore not bound to respect it. The United States had, however, signed the Food Aid Convention of 1999, and the attempt to present Africans with the ultimatum 'Eat GM or starve' was in clear contravention of that measure. The Convention recognises that food aid should be wherever possible locally sourced and avoid distorting markets, and that it should respect people's cultural beliefs and practices. The agreement was part of an attempt to limit the manipulation of food aid which has been, historically, its main feature. As a spokesperson for Oxfam, welcoming the Convention commented, 'Food aid programmes have historically been used inappropriately with industrialised countries using them to dispose of surpluses and create food dependencies.' This is precisely what the US was up to in Africa.[19]

Given that no laws were being broken or treaties breached, and that plenty of non-GM food was available, Zambia's refusal to accept GM food should not have been a problem, even had it not been based on reports from their own scientists and such august bodies as the British Medical Association (BMA).[20] It was a problem purely and only because the agbiotech industry spotted an opportunity to undermine the credibility of anyone harbouring the least reservations. For these would-be dictators of the food supply, it is not enough that we all eat just what they tell us to, we must also be grateful to them for allowing us to do so.

It should also be pointed out that the refusing countries' main concern was that what they were being offered was unmilled maize which could have been planted rather than eaten and which would then have contaminated local varieties, with unforeseeable effects. As a letter from a Namibian resident to the *New Scientist* pointed out, the refusal in some cases to mill this maize before delivery provoked a 'widely spread concern' that food aid was being used as 'an attempt to undermine the viability of the region's agricultural sector by bringing its organic and GM-free status into question, thereby opening up the area to the high-tech multinationals'. As the writer added, though this might be a difficult charge to prove, 'it would not be out of line with the general experience of the region'.[21]

In fact, Zambia's position was well thought-out and perfectly logical. It was not motivated by ignorance, superstition, panic, or any of the other things the country's government was accused of in the western press. Zambian NGOs had made it clear that, given financial support rather than unwanted leftovers from the overfilled plate of the west, they could get food from the parts of Zambia where there was plenty of it to those areas where it was in short supply. Although drought had destroyed much of the maize crop, there was an estimated surplus of 300,000 tons of cassava available in the north of the country, where it was grown as the staple diet. The government's chief scientist, Mwananyanda Mbikusita Lewanika, admitted that the problem was largely organisational: 'We have food', he said, 'but we have no capacity to distribute. We must put our own house in order.' Zambia, moreover, was not offered aid as such, not if one assumes, as most people understandably do, that 'aid' means money or food donated. Zambia received no gifts; instead, Zambian private sector groups received $51 million in loans in order to import GM maize. If the Zambian government is to be believed, moreover, then the US could have given effective help for no more than the cost of sufficient trucks, fuel and Peace Corps volunteers.[22]

It quickly became clear that the US response to Africa's food crisis was part of a wider strategy which none of these alternative approaches would have served. That strategy was already evident at the conference in Rome in June 2002. At the so-called 'World Food Summit: 5 Years Later', the United States Agency for International Development (USAID) announced biotechnology aid worth $100 million, aid which, under the title Collaborative Agriculture Biotechnology Initiative (CABI) was tied to trade commitments and commercial interests. According to Vandana Shiva, the conference

> conveyed the impression of being more a sale-show for the biotechnology industry than a serious gathering of leaders seeking to find collective ways and make collective commitments to address the biggest human rights disaster of our times – more than a billion people going hungry in a world with abundant food and wealth.

Instead of genuinely seeking ways in which the root causes of inequality, poverty and the resulting hunger can be addressed, the World Food Summit was clearly a move towards redefining the terms of reference in which debate around the problem is conducted, 'a redefinition of globalisation as sustainable development and biotechnology as sustainable agriculture'.[23]

This redefinition was spelt out as the Great and the Good gathered once again for one of their now seemingly endless series of useless, expensive summits, this time the 'World Summit on Sustainable Development' – Earth Summit for short. The roots of hunger lie, according to the World Bank and the United Nations World Food Programme, not in imbalances of wealth or power, but in 'economic mismanagement' compounded in some regions by what appears to be seen as sheer bad luck, such as Southern Africa's persistent drought. Anything might be to blame for hunger – in Southern Africa the fault lies with botched land reform, Aids, and presumably God, or whoever it is who decides when it will or will not rain – provided it does not reflect badly on western corporations or the governments whose agenda they increasingly set. The US Farm Bill of 2002, which allocated between $15 billion and $20 billion to subsidies which help maintain an otherwise unsustainable system of intensive agriculture at the expense of Third World farmers, for example, is not seen as 'economic mismanagement' but as a product of unavoidable political realities. Such subsidies are not aimed at restructuring US agriculture but at maintaining it as it is, which means that they keep US farmers in business by encouraging them to over-produce. The results of this over-production end up as subsidised exports to developing countries where they drive local farmers out of business, perpetuating poverty and hunger. With no significant exceptions, commercial development of GMOs has been directed at perpetuating and intensifying this exploitative, wasteful system at the cost not only of Third World farmers but of the environment and the health of US farmers and farmworkers and the consumers they serve. This is why, even in the unlikely event that genetically modified organisms turn out not to provoke serious health problems, they represent a new weapon in what is, even if inadvertently, a genocidal war against the poor.[24]

Genocide – like 'rape' – is not a word to use lightly, though it merely mirrors Bush's statement that EU policies on GMOs were 'letting people starve'.[25] Yet surely the wonderfully named 'United States Leadership Against HIV/AIDS, Tuberculosis and Malaria Act', signed into law in 2003, represented a new low in US foreign policy, one which demands the strongest possible condemnation. The Act linked financial aid to combat serious illnesses – diseases which kill millions – to acceptance of GM food. United States development initiatives have always, invariably, and quite nakedly, been designed to promote the country's own exports. Yet the 2003 Act, which seeks to institutionalise the rape of Africa begun by the Bush Administration's seizing of the opportunity provided by a shortage of food, goes beyond this. Even if the rejection of GMOs were truly motivated not by any real scientific concern but by ignorance and superstition, would such an ultimatum be justified? There is no scientific evidence that eating pig meat is any more dangerous than eating the meat of other farm animals. Yet if mass hunger were to afflict Israel or Saudi Arabia, what would we think of a country which said 'We don't give a damn for your beliefs – eat our surplus pork or starve'?

The Earth Summit turned out, in fact, to be a triumph for African countries in their struggle for the right to determine whether or not they would accept GMOs. A resolution condemning them for refusing such 'aid' was initially resisted only by a number of the African countries themselves and a few small First World allies, notably Switzerland and Norway. However, following an intervention from the Ethiopian delegation, most developing countries, as well as the European Union, swung behind the rejectionists. Neither Ethiopia, nor any other African country, should be forced to accept GMOs in any form, forced, in other words, to solve a short-term problem by inflicting a number of grave long-term problems on itself instead.[26]

The American biotech industry will not, of course, go away. Having only partly succeeded with the big stick, however – and at the price of alerting much of the world to just what they were up to – they will alternate its use with that of a nice, juicy GM carrot. In March, 2003, the Rockefeller Foundation announced that it was establishing an African Agriculture Technology Foundation (AATF) which would 'serve as a platform where African scientists and development experts can access new materials and information on technologies owned by international private companies and later even transfer them into the hands of millions of African farmers'. These 'technologies' in fact consist of just one: genetic engineering. The Foundation states openly that 'Biotechnology's promise of ending hunger and promoting economic growth in developing countries is the main reason behind the AATF.' Its centre in Nairobi was funded not only by the Foundation itself, but by USAID and the UK government, which has committed £13.4 million to promoting GMOs in the Third World.[27]

When the Bush Administration was not blaming the EU for Africa's food supply problems it blamed ignorance and superstition. It is important to remember that this essentially racist portrayal of Africa's reasons for rejecting its food aid has no basis in reality, insofar as many African countries are conducting experimental releases of GMOs as well as laboratory-based research. The attempted rape of Africa had nothing to do with the general view of agricultural biotechnology on the continent, which is as divided as it is anywhere else. What Africans were insisting upon was the right to take all factors into consideration before making their own decision as to what was best to do. The dangers of food aid, and the need to balance immediate need against longer-term considerations, are not exclusively linked to the question of GMOs.[28]

## INDIA

According to a recent Institute of Development Studies (IDS) study of biotechnology in India, 'it is clear that through a combination of material influence, in most cases high levels of institutional access, and in a context in which claims about the benefits of biotechnology are echoed and repeated in influential media, firms have played an important role in the evolving regulatory regime'. The result is that 'the policy agenda in Delhi appears to be far more influenced by a fairly close-knit policy network of biotech entrepreneurs from larger multinationals and successful start-up firms with good national and global connections'.[29]

Both MNCs and indigenous companies with an interest in biotech have been organised since 1994 into the lobbying body All India Biotech Association (AIBA), which has helped to institutionalise the seeking of influence, reducing reliance on traditional informal methods. Corporations also work through the Confederation of Indian Industry (CII) and other trade and business groups. These groups are able not only to bring direct pressure to bear on Indian legislators and regulatory bodies, but to improve collaboration with powerful pro-biotech forces abroad. The IDS study gives the example of a Memorandum of Understanding signed by the CII, the US–India Business Council and the US biotech industry lobby BIO. The agreement covers 'information sharing on trade and investment opportunities in the biotech sector with a particular focus on agriculture, facilitation of "one-to-one" interaction between business and government in the US and India and the establishment of a working group to facilitate trade opportunities'.[30] BIO, the IDS study says 'can provide smaller outfits such as AIBA with resources and materials to support their efforts to represent the industry to government in a positive light'. It is hard to tell here whether the humour is conscious or inadvertent, but judging by the way in which BIO operates on home ground, we might justifiably regard this description of what it has to offer as euphemistic.[31]

### Seeds

This is not to say that India has been subject to the same degree of regulatory capture as is found in the United States. Powerful popular forces, as well as a section of Indian capital and of the scientific intelligentsia, remain sceptical or actively opposed. As the author of the IDS study, Peter Newell, points out:

> given the size of the country and the symbolic weight its actions carry, what happens in India sends out a powerful message to the rest of the developing world. This is one of the reasons that it has become a key site for biotechnology companies and anti-GM activists alike in the global contest over the future of biotechnology in agriculture. The policy debate about the development and regulation of biotechnology in India is therefore strongly affected by global constellations of interests within the scientific, business and NGO communities.[32]

Within India itself 'smaller seed companies...openly question the role of biotechnology in India's agricultural development, raising concerns about impact on seed markets or the suitability of GM crops for the nature of agro-ecological conditions in India'.[33]

The Environmental Protection Act of 1986 (EPA) did nothing to slow the ongoing process of loosening India's regulatory environment as it affects the interests of farmers and those who supply their needs. Two years later a National Seed Development Policy was adopted which allowed firms based in India to import seeds. In 1989, 'Rules for the Manufacture, Use, Import, Export and the Storage of Hazardous Micro-organisms, Genetically Engineered Organisms or Cells' were introduced as an annex to the EPA. Under these rules, a Genetic Engineering Approval Committee (GEAC) was established whose permission was required before GMOs could be imported or released into the environment. A separate body, the Review Committee on

Genetic Manipulation (RCGM) issues permits for experiments which are confined to laboratories, sealed greenhouses or other closed environments.[34]

The immediate reaction of foreign corporations to the new regulatory environment was to acquire Indian concerns which, unlike them, were allowed to import GM seeds. General economic liberalisation in the 1990s provided an increasingly conducive environment for MNCs' activities in the country. Licensing in the seed sector was abolished and collaboration between Indian and foreign firms actively encouraged. Investment in the sector from private sources rocketed, more than tripling between 1993 and 1997. At the same time, however, external trade in staples remained under strict control, severely limiting the possibilities open to biotech companies by removing wheat, rice, soya oil and cottonseed oil from the picture. Legislation specific to the cultivation and marketing of GMOs was first enacted in 1990 and updated further in 1994, 1998 and 1999. Releases, whether commercial or experimental, must be approved by the GEAC. This is by no means a routine matter, and GEAC has turned down numerous applications, exercising particular caution in relation to commercial releases. However, as the GEAC does not have the power to punish offenders, complaints that its decisions have been flouted must be heard in the courts under the EPA, and even when such complaints were upheld the punishments meted out have been derisory. In 2001, a National Seed Policy was adopted, which according to Newell 'seems set to consolidate this pattern of growth within a liberalised economy'.[35]

#### The influence of foreign corporations on the Indian government

Increasing collaboration between Indian firms and biotech MNCs such as Monsanto has also improved the multinationals' access to political decision-makers. Newell gives the example of Monsanto's purchase of a 26 per cent holding in the Maharastra Hybrid Seed Company (MAHYCO). This was seen as a wise acquisition because

> Mahyco's director Dr Barwale is a well respected member of the Indian agricultural industry who has been honoured by the Indian government...His connections within government extend beyond the Department of Biotechnology [DBT] to many of the key agencies involved in biosafety regulation.[36]

The additional influence such connections give to Monsanto undoubtedly helped it to win the first approval of a commercial release, granted in March 2002. The permit, for Bt cotton, covered six states and was hedged in by a number of conditions, but its granting represented a breakthrough victory for biotech in industry and a vindication of Monsanto's expensive lobbying strategy, which included maintaining a permanent office of 'regulatory affairs' in Delhi. In addition, the company has engaged in a major public relations onslaught using videos, newspaper ads, open days at its research laboratories, workshops and opinion surveys. This campaign has even been berated by rival biotech MNCs and others who see GMOs as having the potential to improve India's agriculture as 'unnecessarily aggressive and insensitive' and responsible for 'doing biotechnology a great damage...set[ting] things back by one decade'.[37]

According to the Indian scientist and environmental activist Vandana Shiva, Monsanto did not win its permission to import and eventually market Bt cotton seeds only by winning hearts and minds through sustained propaganda. It did so by persuading the state to break its own laws. 'From beginning to end the introduction of genetically engineered Bt cotton in India has been illegal,' Shiva has written. 'Bt genes were imported illegally by MAHYCO without GEAC approval. In 1998...[Monsanto and MAHYCO] started 40 open field trials in 9 states without GEAC approval...in violation of the Biosafety laws.' Illegal trials continued the following year in ten locations. 'They continued,' Shiva says, 'to multiply seeds in total contravention of the Bio-safety laws which require that all planting material at trial stage...be destroyed.' She lists further violations of the law, arguing that these have been so widespread that they have resulted in the contamination of 10,000 acres in Gujarat, a state with '130 indigenous cotton varieties'. Given that the approval of Bt cotton for India was widely trumpeted in the western press as 'a breakthrough for the biotechnology industry in the developing world', the fact that the Indian authorities broke their own laws, and the fact that this was successfully exposed, could have serious implications for the United States' prospects of success in their attempts to use the Third World as a dumping ground for its unwanted GMOs and the unwanted technology which produces them.[38]

### The public sector

While the private sector subverts the law and conducts a huge public relations/propaganda campaign, Indian public sector institutions are in the process of doing transgenic research into a number of food crops, including rice, mustard, potato, tomato and brassicas, as well as non-food crops, principally cotton and tobacco. Much of this research is financed from overseas sources, or involves private sector/public sector collaboration and concentrates on 'second generation' projects designed to improve the nutritional qualities of common foods. A typical example is the joint project conducted by the Tata Energy Research Institute in Delhi, Monsanto, Michigan State University and USAID to develop a variety of GM mustard that will yield cooking oil high in beta-carotene. Other projects, however, focus on the usual commercial target, pest resistance.[39]

#### Illegal planting of GMOs

Although approvals of releases remain relatively limited, it is well known that, as in Brazil, GM crops have been illegally grown for the market. In Gujarat, a state which has close to 2 million hectares under cotton cultivation, complaints were received by the government in 2001 that several farmers in the state were cultivating Bt cotton. Ironically, the principal complainant was a seed company which had been conducting large-scale field trials on Bt cotton since 1998 and was awaiting the approval for commercial release. Investigations then revealed that the illegal cultivation covered close to 10,000 hectares. Other cultivations were discovered in the wake of these findings, though the acreage involved was relatively small.[40]

The discovery that some farmers were routinely violating what were on paper strict biosafety regulations provoked the government, as well as environmentalists and small farmers' organisations, into action. Illegal cultivation called the adequacy of these regulations into question.[41]

#### Resistance – and disinformation

Delays to commercial approvals have come about largely through massive pressure from one of the world's biggest environmentalist movements.[42] Popular resistance and criticism from scientists even forced Syngenta to abandon a joint research project with the Indira Gandhi Agricultural University (IGAU) which would have given the firm commercial rights to over 19,000 strains of local rice cultivar stored there.[43] However, the environmental movement, which embraces urban intellectuals, NGOs such as Greenpeace India, and the vigorous organisations of small farmers, is up against a disinformation campaign that in particularly high gear to pursue the commercial imperative of inflicting GMOs on the South.

Take the case of the Bt Cotton which, according to a headline in the *New Scientist,* passed a ten-year trial 'with flying colours'. The GM cotton 'cuts the numbers of a devastating moth pest and could eventually eliminate it...' Researchers 'found that numbers of the pest fell in areas where 65 per cent of the cotton fields were growing the Bt variety', concluding on the basis of 'mathematical models' that 'if farmers planted 80 per cent of their fields with Bt cotton they would eradicate the pest completely in several years'. [44] Others saw things differently. Bt cotton was supposed, for example, to be resistant to the pest species bollworm, but experience following the first commercial releases showed that this was not the case. According to one account, Bt Cotton was completely wiped out in three states, attacked not only by bollworm but by pests and diseases hitherto unknown in India. GEAC, and Indian farmers, had been hoodwinked. When farmers harvested their first crop of Bt cotton at the end of the 2002 growing season, many discovered that the seed for which they had paid up to four times the going rate for non-GM cotton yielded poorly in quality and quantity. Greenpeace openly accused the government of lying and demanded an investigation. It was, moreover, not only disgruntled farmers who complained. The seed breeders' association also noted that the variety had 'failed to give expected results', though they claimed that there was no problem with Bt cotton *per se*. And some farmers did much more than complain. In September, 2003, a group of around forty *ryot* stormed a Monsanto research facility in Bangalore demanding that the firm leave India, blaming them, amongst other things, for a wave of suicides amongst farmers ruined by bad weather but also the false promise of GM cotton. This was, moreover, by no means an isolated incident, but one of a series of such disturbances.[45] Whether the failure of GM cotton was inevitable is unclear, as MAHYCO had simply chosen the wrong kind of Bt cotton for Indian conditions.[46] The authorities had also been negligent in granting approval on the basis of blatantly inadequate evidence, a 'mistake' which they would repeat, notably in approving a GM mustard despite evidence of inferior yields and serious irregularities in experimental trials purportedly designed to assess the crop for safety.[47]

These experiences may have helped to underpin the continuing caution in India evident at the beginning of 2003 when the country's authorities refused a shipment of food aid from the US on the grounds that it might contain GMOs. Once again, the real reason for the massive global propaganda campaign on behalf of this technology and its products was revealed: having invested billions of dollars in research, development and production of genetically modified food, they can't even give the stuff away.[48]

## ARGENTINA

Argentina is a country which has stood, in the recent past, on the verge of prosperity. In the 1960s its GDP equalled that of Italy, before a mixture of thuggish incompetence in government at home and exploitative manipulation from abroad gradually weakened the country to the point where it became vulnerable to the economic collapse which overtook it at the dawn of the twenty-first century. Argentina has thus gone, in the space of a couple of decades, from amongst the most prosperous parts of Latin America – a country which was often described as 'the bread basket of the world' – to complete economic collapse. Guided by neoliberal economics imposed by the United States and the international agencies it dominates, its people have lost everything: those who were middle class have become impoverished, and those who were already poor stand on the brink of catastrophe. From bread basket to basket case, and in the time it takes a baby to grow into an adult. It has also become what one expert describes as 'among the more advanced developing countries in terms of current and intended use of genetically engineered crops and products derived from them'. These two facts are not unrelated.[49]

The decline of the industrial sector, persistent financial and political crises, and the utter prostration of the authorities in the face of disastrous economic policies imposed by the International Monetary Fund (IMF) made the country peculiarly vulnerable to outside pressure. Over the last quarter of the twentieth century Argentina's economy was restructured. From having a mixed agricultural-industrial base, it was transformed into a much more typical 'Third World' country, dependent not merely on agriculture but on a single agricultural commodity, soya beans. After 1970, when soya beans occupied only 38,000 hectares of Argentina's agricultural land, production steadily increased to an estimated total for 2004 of 15 million hectares. Most of this crop is converted to oil for export, accounting for over four-fifths of the world's soya oil and over a third of its soya bean meal, products which are found in a huge range of processed foods. In the 1990s, deindustrialisation was worsened by the removal of tariff barriers, which led to a flood of imported goods with which indigenous production could not compete.

This huge growth can be directly attributed to the policies imposed on the country by the IMF. The IMF's prize pupil, Argentina's President Carlos Menem, little more than an IMF puppet, signed contracts with Monsanto and Cargill early in the 1990s, contracts which set the country firmly on a course away from diverse agriculture and towards soya monoculture. Almost all of this soya is genetically modified, yet despite this potentially controversial aspect no debate took place either in parliament or in the streets, Menem signing his country up to the GM wonderland without bothering to inform the citizens or their representatives.

The IMF's Structural Adjustment Programme encouraged the conversion of pasture to arable land, with the emphasis increasingly on soya and maize for export. The result has been a devastating depopulation of the countryside:

> Migration to the cities has risen at an alarming rate: 300,000 farmers have deserted the countryside and more than 500 villages have been abandoned, or are on the road to disappearance. Agribusiness GM soya farming requires agriculture without culture or people. As a consequence, the *villas miseria* on the outskirts of the cities are mushrooming with the arriving unemployed agricultural workers.[50]

Tax reform attracted major foreign investment, including by Monsanto, who chose Argentina as an ideal location for the expansion of herbicide production. High agricultural prices in 1997 further contributed to the spread of soya, including – and increasingly dominated by – GM varieties. By this time, however, the national debt, which had dwindled to almost nothing in the early part of the decade, was rapidly growing into an unsustainable burden. As one analyst explained,

> [these] circumstances enhanced the need to further exploit the natural resources and reduced the importance of industrial activities which could add value to the raw materials produced. Thus Argentina found itself exporting leather and importing shoes; exporting cotton and importing textiles; exporting cereals and importing pasta and biscuits, exporting gas and petroleum and increasing its deficit in petrochemicals and refined oils. As a result, the balance of trade deteriorated and the debt crisis deepened.[51]

Argentina began exporting GMOs commercially in 1996, when the Argentina Commission of Assessment in Biotechnology (CONABIA) granted its first approval, licensing Monsanto to place Roundup Ready Soya on the market. In order to win a significant slice of sales, Monsanto offered very favourable prices for Roundup Ready soya seeds and the glyphosate which went with them. Argentina's regulatory system was thus designed almost from the outset to be as corporate-friendly as possible. Guidelines supposed to ensure the environmental safety of GMO releases were based on those operating in the US, and so were minimal and largely cosmetic. Existing laws were adapted to encompass the application, review and approval procedures meant to guarantee the safety of foods. GM seed registration was also conducted under the pre-existing system.[52]

Not content with relying on exports, the GM soya industry has attempted to persuade Argentines to eat them. In this case, in contrast to what is occurring in Africa, seduction has been preferred to rape:

> Soja Solidaria encourages soya producers to donate 1 per cent of their soya production to comedores – eating halls for the unemployed, and in public schools, hospitals, neighbourhood centres and old people's homes. The organisation uses community participation to reach the heart of society, complementing their donations with cooking courses using soya recipes and the provision of health and nutritional advice on the benefits of the genetically modified bean.

Labelling of GM products is not required in Argentina, so neither these lucky recipients of charity nor paying customers have any way of knowing whether what they are eating contains genetically modified organisms or products derived from them.[53]

All of these developments occurred under the aegis of the Ministry of Agriculture, with the Ministry of Environment sidelined. This was bad enough, but in 2002 the Ministry of Agriculture was replaced by a new department covering agriculture, livestock and fisheries. Bizarrely, a country now almost entirely dependent on an ever-more intensive agriculture for its survival no longer had a fully fledged ministry to deal with the sector, the new department being part of the Ministry of Finance, though food safety evaluations continue to be conducted through the Ministry of Health, whose advisory committees conduct technical reviews and make, or withhold, recommendations for approval of individual release applications.

The new department of the Ministry of Finance, however, continues to have the last word on field tests and commercial releases. As a consequence, regulatory systems have gone from weak to risible. A shortage of expertise forces the country to rely on advice from the only people around who know anything about the subject, people generally compromised by their relationships with the firms whose products they are supposed to be assessing.[54]

Ignoring the environmental consequences not only of the GMOs but of the monoculture they are designed to encourage has had devastating consequences:

> Floods without precedence are taking place as forests are cut down to make way for soya crops. In the high-mountain provinces of Salta and Juyuy, on the border of Bolivia, the subtropical Yungas region is being deforested to make space for soya plantations. Greenpeace has warned that in five years, the ancient cloud forest will be extinct.[55]

Argentina now takes over a quarter of its total income from exports from soya oils and seed exports. The government is aware of the possible problems attendant upon this and has proposed increasing taxes on GM exports, enabling farmers to choose other crops. The increasing reluctance on the part of the EU and much of the rest of the world to import GMO-based products may yet leave Argentina in the lurch, a predicament imposed upon it by the short-sighted, socially and environmentally destructive policies favoured by the IMF.[56]

The problem is neatly summarised by researcher Joel I. Cohen who, in an article in the *Journal of Human Development*, noted that

> Biosafety evaluations on Argentina…focus on risk in a proposed release. The task is to identify any potential risk and explore potential means for managing identified risks. Ostensibly, evaluations compare predicted impacts of the GMO with those of the equivalent non-GMO variety. Genetically modified varieties that present no greater risk than the referenced conventional variety are deemed acceptable for testing and eventual commercial release [a system very similar to that of the US]
>
> …
>
> [Because] the impetus for building biosafety infrastructure came in the form of external requests from transnational companies seeking a place

> to grow bulk commodities and for off-season seed production...the country's introduction to GMOs was in the form of imported transgenic soybean and maize varieties grown almost exclusively for export.

The result is that the

> potential for conflict of interest is an inherent part of Argentina's biosafety system. Nearly all biosafety reviewers conduct applied research at public institutions (leading to field tests, and possibly commercial products), work collaboratively with biotechnology companies, or belong to industry organizations...The prevalence of these relationships makes it common for a Commission member to excuse himself from taking part in a decision. Such connections also make it difficult to find independent, disinterested members to review applications containing confidential business information.[57]

## BRAZIL

Brazil's course in relation to agricultural biotechnology has been quite different from that of its southern neighbour. In principle, growing GMOs for the market has never been permitted, while experimental plantations have been strictly controlled. In those areas bordering neighbouring countries where cultivation is legal and widespread, however, MNCs dependent on agricultural biotechnology have deliberately encouraged illegal planting of their seed, with catastrophic effects in limited localities. In an attempt to persuade the Brazilian authorities that GMOs were the irresistible tide of the future, the biotech propaganda machine has created the entirely false impression that contamination is now ubiquitous and the game therefore up for GM-free agriculture. This had a precedent in the decision in 2002 by Pakistan to lift its ban on GM imports, ostensibly because contamination had become so widespread through illegal cultivation that there was no longer any point in trying to keep GMOs out.[58] Whatever the truth of the situation in Pakistan, that this is not the case in Brazil is confirmed by the private company Cert-ID, which makes a living by tracing, amongst other things, non-GM products from field to fork. According to Cert-ID's Augusto Freire, this is true of only the extreme south of the country. In many states there are no GMOs at all, and in the great food-producing central states contamination from experimental plantations has so far been avoided.[59]

This is important because Brazil, currently second in the league table, is poised to overtake the United States as the world's biggest producer of soya beans by 2006. Were it not to grow GM soya, it would not face the difficult problem of segregation and would have a ready market for its goods, particularly in Europe. In recent times, demand for non-GM soya has seen exports to Europe and Japan increase dramatically, while allowing growers to charge premium prices.[60] The conflict is therefore not entirely, or even mainly, to do with arguments about the possible dangers of GMOs for the environment or public health, but a straight commercial fight between those who wish to grow GMOs and those whose livelihoods depend on their being able to supply a guaranteed GM-free product.[61] In the background, as ever in Latin America, squats the US and its profit-hungry corporations. As the *Financial Times* reported, Monsanto and others, aided by the US government, have launched a major effort to win hearts and minds in Brazil, paying for visits to the US by Brazilian scientists and politicians, as well as attempting to win over the broader public through advertising.[62]

With the election of progressive president Luíz Inacio Lula da Silva in 2002, the rules of the Brazilian game changed. The left's victory might have been seen as a setback by Monsanto and the other MNCs seeking to engineer a GM takeover of Brazilian agriculture, but these people are nothing if not resiliently optimistic. Lula had promised to launch a sustained campaign to eradicate hunger in his country. Monsanto, of course, was there to help. As Monsanto Chairman Frank AtLee said, 'The new government has talked about being opposed to biotech. However, it has a major interest in feeding the hungry people of the country. This is a very dynamic situation. We are working very hard with the government to demonstrate our products can help them.' Monsanto, losing money and virtually ejected from the EU, was desperate for a breakthrough elsewhere to cheer its long-suffering shareholders.[63]

Things did not, at this stage, look too promising for Monsanto. Rio Grande do Sul was the largest of a number of states which had already declared themselves to be GM-free zones, and the issue had become a major element in the election campaigns and political profiles of an increasingly popular left and green movement. Lula's agricultural policy adviser during the election campaign, José Graziano da Silva was on record as saying that their goal was 'to establish a reputation as GM-free'. The president's party, the Workers' Party (PT), had a large rural following which was generally hostile to agricultural biotechnology and its products. Moreover, the predecessor right-wing government had made several attempts to end the ban and had at each turn been thwarted by successful court actions brought by consumer and environmentalist groups. In addition, the message seemed to be spreading, with more Brazilian-based food companies declaring that they would no longer use GM ingredients. Perdigao, for example, which trades in both animal products and animal feed, stated its intention to eliminate GMOs from its operations completely.[64]

What may turn out to be a breakthrough for Monsanto and its allies occurred at the beginning of the planting season in September, 2003, when the government which had pledged to stand firm against GMOs found itself in the middle of an undignified public squabble over the issue. In June, Lula's Chief of Staff, José Dirceu, had reiterated the promise not to allow cultivation of GMOs. Three months later, however, while Lula was on a visit to Washington DC, Dirceu recanted, saying that permission would be granted. To everyone's surprise, the Vice President José Alencar – not a member of the Workers' Party – who, under the Brazilian Constitution enjoys full executive power during the president's absence abroad, said he could not approve the measure, which he argued was illegal. Lula was then quoted in the US press as saying, somewhat enigmatically, that Alencar 'knows what he has to do and will do it'. Alencar, who perhaps suspected he had been set up to take the blame for an unpopular measure, eventually backed down and signed the 'provisional decree' allowing commercial planting. This was, however, far from being a carte blanche. It applies only to approved varieties of one crop – soya beans; it contains severe geographic restrictions, keeping GM plants away from non-GM plantations, nature reserves and water courses; it forbids GM seeds from being transported across state lines; and it obliges anyone wishing to grow GMOs to agree to pay compensation for any damage to the environment or public health.[65]

Despite these restrictions, the decree clearly reneges on an election promise and almost certainly represents the thin end of a potentially disastrous wedge. Brazil is risking not only the health of its people and their environment, but the loss of the huge commercial advantage of being able to guarantee GM-free crops. As the *New York Times* reported,

> The government's about-face is...likely to provoke tensions in the hitherto warm relations between Mr. da Silva and his allies and admirers in the Green movement in Europe. His Workers' Party has been the main

> sponsor of the annual World Social Forum in the southern state of Rio Grande do Sul, which has emerged as a magnet for anti-globalization groups whose agenda includes strong opposition to genetically modified products.[66]

Reaction to the apparent betrayal both within and outside Brazil was furious. Greenpeace, farmers' group Via Campesina, and many others had been converging on Brasilia since early September to protest and attempt to prevent the impending announcement. Most embarrassingly for Lula, Greenpeace was able to produce a questionnaire it had sent to presidential candidates with clear proof that Lula had unequivocally pledged not to allow commercial growth of GM crops.[67]

## ISNAR AND OTHERS: THE CARROT TO BUSH'S STICK

In the developing world itself, attitudes amongst the general public as well as decision-makers are as varied as they are in industrialised countries, and generalisations are not possible. In many developing countries, biotechnology is seen positively, at least by governments and other powerful forces. Biotechnology research and development in the South tends to lean heavily on the public sector. By providing assistance for capacity building and specific projects in developing countries, the industrialised world has been able to push more countries in a pro-biotech direction in a way which seems likely to enjoy more lasting success than will bullying governments with hungry people to feed.

In this respect the capacity building assistance given by the International Service for National Agricultural Research (ISNAR)[68] plays a crucial role. ISNAR's first biotechnology oriented project, 'Agricultural Biotechnology: Opportunities for International Development', began in 1988 under the additional co-sponsorship of the World Bank and the Australian government. On the basis of this project's findings, ISNAR then established the Intermediary Biotechnology Service (IBS) which now co-ordinates its work in this field, its aim being 'to provide national agricultural research systems [NARS] with ready access to information and impartial advice on the best ways to use modern biotechnology to solve agricultural problems' as well as 'how to integrate biotechnology into existing agricultural research programs'. Funding for the IBS comes from the governments of the Netherlands and Switzerland through their aid and co-operation programmes, with some further funds coming from state agencies in Sweden, Japan and the OECD Development Centre.[69]

ISNAR urges governments to adopt 'a clear policy and agenda for biotechnology research'. The key to capacity building is the forging of 'partnerships between developing-country research systems, international research institutions, and private and public sector research organizations in industrialized countries'. In addition, governments are encouraged to 'provide incentives for the private sector to undertake biotechnology research that focuses on farmers' problems'. ISNAR's own programme of research, moreover, focuses on the elements necessary for successful capacity building, as well as analysing the impact of biotechnological innovations and interventions, setting priorities, the impact on developing countries' agriculture of differing frameworks for the protection of Intellectual Property Rights (IPR) and biosafety.[70]

Capacity building is a tempting notion for developing countries used to seeing the few scientists and managerial personnel which they are able to train heading north at the earliest opportunity. Even those with little enthusiasm for agricultural biotechnology *per se* may be attracted by the highly transferable skills they are likely to acquire. ISNAR stresses such aspects, along with a need to link, as one ISNAR publication puts it 'policy, capacity and regulation'.[71] And ISNAR, despite its claim to neutrality, does not see the use of modern biotechnology in agriculture as, *per se*, a contentious issue. For ISNAR, the only contentious issues concern the means by which such technologies can best be harnessed for the good of poorer countries and poor people. Its focus is therefore upon giving developing countries the means to join the game, which, before we even arrive at capacity building, means convincing them that it is a game worth playing. Crude propaganda and downright threats play a major role in this, as we have seen, with capacity building and an emphasis on biosafety as the Mr Nice to such Mr Nasties as the Bush Administration's attempted rape of African agriculture.

ISNAR both advocates and assists with the development of national biosafety systems which 'integrate political, social, ethical, health, economic, and environmental considerations into decisions regarding the safe and appropriate use of biotechnology methods and products'. Its strategy is to convince developing countries' decision-makers, as well as influential people in the industrialised world, that, far from threatening to reinforce dependence and its attendant inequalities, biotechnology in fact provides the key to sustainable development. Unlike corporate spokespeople, ISNAR admits that enhancing agricultural productivity cannot in itself reduce hunger or provide better livelihoods for more people. Nevertheless, the group argues that 'while only part of a total solution that involves better market conditions, policies, and access to production resources, biotechnology can contribute to addressing poverty issues in developing countries'. It purports to a degree of neutrality, admitting 'a pressing need to document both the positive and the negative effects of biotechnology in rural communities' as part of the 'ongoing debate about biotechnology'. Yet it is clear from the material it produces and the programmes it helps fund that, for ISNAR, this debate is not about 'whether' but about 'how'. Biotechnology is around, it works, it can be made safe. The only legitimate criticism, for ISNAR, is that 'poor farmers and consumers stand to benefit very little...' The task, then, is not to investigate alternative technologies which might carry such benefits, but to see how this one can be turned to the general good.[72]

ISNAR claims to work according to what it calls the 'Sustainable Livelihoods Framework', which 'draws on many considerations often excluded from agricultural economic impact studies', and works with farmers themselves to improve their lot. A consultation exercise conducted by ISNAR during 2001 identified what it described as 'several unique characteristics of biotechnology', an approach which again assumes that biotechnology, though it may be attended by problems, is essentially part of the solution rather than part of the problem. ISNAR's pro-biotech strategy is to accept many of biotechnology's opponents arguments but reject their conclusion, arguing instead that such matters as the danger that biotechnology will reinforce divisive market structures, or that it will not benefit those people most in need, that it puts the environment, particularly biodiversity, at risk and may pose a threat to human health can and should be addressed. Far from threatening to reinforce the inequalities apparent in the functioning of global markets, if applied correctly 'Recent advances in agricultural applications of modern biotechnology show a significant potential...to contribute to sustainable gains in agricultural productivity, reducing poverty, and enhancing food security in developing countries.'[73]

## THE EUROPEAN COMMISSION: PRO-BIOTECH PROPAGANDA

The same message can be found in material produced by a range of official and semi-official bodies. In *Creating Sustainable Solutions in Developing Countries*, for example, the European Commission reports GMO field trials alongside projects which use remote-sensing techniques to improve the sustainability of forest resource exploitation, the creation of a global database on fish biodiversity, and the development of more ecologically friendly systems of aquaculture. In one, 'scientists made genetic improvements to maize plants' in order to enable them to be cultivated in soils too acidic for conventional maize, drawing the conclusion that 'high productivity and sustainability can only be achieved by both genetic and agronomic strategies...'[74]

One of the most naked acts of pro-biotech propaganda conducted by the European Commission was the supposed forum for discussion of development and agricultural biotechnology that it hosted in January 2003. Under the guise of an open debate, the Commission used €0.5 million of taxpayers' money to produce two days of skilfully stage-managed 'debate' designed to give the impression that (a) African farmers are desperate to be allowed to grow GMOs and (b) only three things are stopping them from doing so: the unscientific, ignorant, 'Luddite' reaction of 'Greens' from privileged, western, urban backgrounds; their unholy alliance with various First World economic interests; and – just to show that our criticisms are even-handed – the failure of biotechnology companies to address the problems of poor farmers by researching such things as drought- and salinity-resistance, micronutrient content and the management of pests which do not also happen to afflict rich farmers in the North. The climax of this exemplary exercise in bogus consultation came at the end, when an obnoxious individual who appeared to have been recruited from low-budget daytime TV passed amongst the audience with a microphone, blatantly removing it if the speaker questioned the accepted wisdom. When a representative of Greenpeace pointed out that one of the enthusiasts for GMOs who had spoken was the head of an institute funded by biotech MNCs, the comment was treated as if – the scientist in question being female – it were some kind of breach of gentlemanly etiquette rather than a point of vital significance.[75]

## OTHER ADVOCATES OF BIOTECH AS A SOLUTION TO THE PROBLEMS OF DEVELOPMENT

The Rockefeller Foundation and its Consultative Group on International Agricultural Research (CGIAR), the Nuffield Council on Bioethics, the (British) Royal Society and numerous other research groups, foundations and think tanks, some – though not all – of them funded in part by major biotech interests, present similar arguments. Unlike the strident and plainly self-serving nonsense spouted by US trade representatives and politicians, material produced by these groups tends to be nuanced and, to some degree, balanced. The argument tends to be that GMOs, conventional crops and methods, organics, integrated crop management and whatever else may come to hand all have a role to play in solving the problem of food supply.[76] The Rockefeller Foundation's Madison Initiative, for example, does not argue that GMOs can solve the long-term problem of food supply, but that they are needed to make up for the shortage of labour created by the Aids crisis. In addition, the Foundation financed the International Program on Rice Biotechnology (IPRB), which ran for the last 15 years of the twentieth century and cost $105 million. According to one critic, the IPRB was largely responsible for the turn to biotechnology of many Far Eastern developing countries, its 'fruitful legacies' including 'China, India and Korea firmly integrating biotech into national rice research programmes, and the Philippines, Thailand and Vietnam all moving in that direction'. In 1999, on the basis of this work, the Foundation established its 'New Rice for Africa' (NERICA) programme to develop GMOs 'suitable' for conditions found on that continent.[77]

The Rockefeller Foundation is not alone in promoting biotechnology in developing countries through providing funds for research and development. Backed by aid from, *inter alia*, Australia, Canada, the EU, New Zealand and Japan, the Sub-Committee on Biotechnology of the Association of South East Asian Nations' (ASEAN) Committee on Science and Technology (COST) is merely the most important of a number of official ASEAN bodies promoting both agricultural and non-agricultural biotech projects in the region. Beyond ASEAN, the Asia-Pacific International Molecular Biology Network, USAID, the Food and Agriculture Organisation (FAO) and individual countries have provided aid or investment which is effectively designed to make it impossible to allow policy makers to choose any development option other than that which involves the development of biotechnologies.[78]

Richard Hindmarsh, from whose research this list of actors is taken, has been led by his investigation of these funding sources to the conclusion that the function of these bodies – aided by the World Bank – is to help to 'incubate the foundations for molecular empires and pave the way for GM giants, especially through national agricultural research programmes'. While helping to keep scientific talent from emigrating, and providing training in useful and transferable skills, they effectively close off alternative policy options, drawing developing countries along the biotech path, usurping the right of these countries and their peoples to make their own decisions regarding, for example, the best way to maintain and enhance food production. A lack of consultation with these affected populations is a ubiquitous feature of the behaviour of these funding bodies and the research institutes and other facilities which they make possible. Regulation therefore develops increasingly in the direction of the creation of an environment conducive to the health of MNCs rather than that of farmers or those who eat their produce. As Hindmarsh notes, such 'regulation has to be seen to be effective and accountable' in order to counter the rise of anti-biotech movements and 'foster public acceptance'.[79]

This is precisely what was done in the European Union, with the difference that the balance of forces within Europe is less overwhelmingly favourable to biotech MNCs and their interests, so that the legislative framework governing the release and marketing of GMOs is more balanced and does offer the possibility of effective protection of consumers and the environment. In the South, the odds are stacked much more heavily in favour of the multinationals. This is why the EU's decision to introduce a regime much more rigorous than the non-regulation prevalent in North America is of such importance. US claims that GMOs were 'substantially equivalent' and posed no particular problems were undermined when the world's largest market economy decided to reject the same approach, one which it had originally been inclined to emulate. As Juan Lopez of Friends of the Earth Europe has argued, a number of factors, including 'the strong presence of consumers' rights groups with considerable lobbying power, the popularity of organic foods, the climate of fear surrounding issues of food safety, and the significance of the European market for agro-business and huge biotech research investments', make the EU crucial to the future of the biotech industry.[80]

## THIRD WORLD SCEPTICS

Stick or carrot, many Third World scientists and the governments for which they sometimes work remain unconvinced. As Eusebius J. Mukhwana of Kenya's Sustainable Agriculture Centre for Research, Extension and Development in Africa (SACRED) explains,

> biotechnology has produced plants that are salt, pest and even drought resistant. These are grand developments by any measure. However, these advancements, while laudable, cannot go far in reducing the poverty and hunger that afflicts this continent...because this poverty is structurally rooted in an unfair and exploitative system of international trade and resource control.

Even if yields could be improved as the pro-biotech propaganda claims, on its own 'this cannot improve the lives of the poor in rural Africa as storage, transportation, marketing, distribution and ability to purchase the food remain nagging problems'. Rather than inadequate yields, what lies at the root of the problem is that the 'best lands and resources are committed to producing cash crops for export, leaving production of staple foods to poor and marginal areas, so that 'people who are dying from hunger are exporting coffee, tea or cotton', a situation which 'biotechnology is most unlikely to change'. Mukhwana is no opponent of biotechnology in agriculture, believing that it 'can help improve the situation'. He is merely alert to the danger that the growing emphasis on biotechnology as the 'solution' to the 'problem of feeding the world' is distracting people from the real problem and the real tasks involved in solving it.[81]

The 'mix-and-match' approach – taking the best ideas from all available technologies, whether conventional, organic or GMO-based – sounds reasonable and attractive. It would be, were the different approaches not in reality incompatible. This is not simply because, for many crop species, co-existence of GM and non-GM crops is impossible to manage, or at the very least extremely difficult. More fundamentally, it is because agricultural biotechnology represents a further step in the development of the kind of farming which has proved environmentally ruinous and incapable of delivering wholesome food. This is an agriculture based on the idea of a technological solution – chemical fertilisers and pest control agents, genetic manipulation, technologies which require constant inputs from outside the farm and its immediate environment – to any problem limiting production.

## THE REAL SOLUTION LIES ELSEWHERE

The only possible long-term solution to the problem of persistent hunger in developing countries must address the fundamental issues of food sovereignty, political and economic independence and participatory democracy, environmental degradation and poverty which lie at its roots. Too much land in the South is devoted to the production of food for export. This is not to say that crops for export cannot provide valuable income to poor countries and poor people. If they are to do so, however, and in a way which will help erode rather than perpetuate hunger, those countries and people must be freed from the huge imbalance of power which burdens their attempts to participate in the market, rather than simply being its slave. This means enabling people to organise on a social and economic level, enabling them, in the words of Miguel Altieri, a professor of environmental science at the University of California at Berkeley,

> to attain food security and conserve the ecological integrity of their farms...If you don't empower them in this way and go right into export agriculture, like many small farmers have done with coffee in Latin America, they become subject to the vagaries of markets.[82]

## FURTHER READING


Vandana Shiva *Monocultures of the Mind: Perspectives on biodiversity and biotechnology* (London/New York/Penang: Zed Books/Third World Network, 1993)

Vanadana Shiva *Protect or Plunder: Understanding intellectual property rights* (London/New York: Zed Books, 2001)

Jonathan King and Doreen Stabinsky, 'Biotechnology under globalisation: the corporate expropriation of plant, animal and microbial species', *Race and Class*, Vol. 40, Nos 2/3, October 1998-March 1999, pp.73–90

*The Corner House Briefing 10: Food? Health? Hope? Genetic Engineering and World Hunger* (Sturminster Newton: The Corner House, October 1998)

Robert Ali Brac de la Perriere and Frank Seuret, *Brave New Seeds: The threat of GM crops to farmers* (London and New York: Zed Books, 2000)

European Commission *Towards Sustainable Agriculture for Developing Countries: Options from life sciences and biotechnology* (European Commission Community Research, 2002)

*Feeding or Fooling the World? Can GM really feed the hungry?* (Five Year Freeze, 2002), <www.fiveyearfreeze.org/Feed_Fool_World.pdf>

Devlin Kuyek et al., *Blast, Biotech and Big Business* (Philippines: Biothai/GRAIN/KMP/MASIPAG/PAN Indonesia/Philippine Greens/UBINIG/Swedish Society for Nature Conservation, 2000), <www.grain.org/docs/blast.pdf>

*Engineering Nutrition: GM crops for global justice* (Food Ethics Council, 2003)

# 5
# International Treaties and Agreements

The regulation of trade in food and other agricultural commodities derived from biotechnology requires the updating of a long-standing system of health and safety protection on the international level. Although the technologies and organisms involved are novel, the problems associated with their regulation are familiar. Depending upon their specific responsibilities and their own ideological position, regulators and those who seek to influence their work will ask of GM products the same questions that they would ask of other commodities: are they safe to eat or use? what is the environmental and social cost of their production? do they undermine in any way the accepted principles of an increasingly liberalised market? and how might their trade be effectively supervised? On the other hand, genetic engineering may also in some cases raise ethical problems, only some of which will be familiar, while away from agriculture, in such fields as medicine and assisted reproduction, ethical debates dominate, presenting even honest and well-intentioned legislators with difficult problems, problems which often arouse heated passions on all sides. Moreover, GMOs as food present difficulties which require urgent solution. Are they safe? If so, under what conditions should they be cultivated and processed?

These simple questions should really have been answered before GMOs were allowed to cross borders at all, as traded items or for any other reason. We have seen, in the attitudes and behaviour of the United States, why they were not. The development and spread of GMOs is not a natural or 'inevitable' process which reflects the superiority and attractiveness of the technology and its products. On the contrary, it is the result of the aggressive and unscrupulous methods employed not only directly by the US and its corporations, but in some cases by other rich countries, as well as by the international financial institutions through which they exercise their will. The World Bank, for example, has been an enthusiastic supporter of the idea that agricultural biotechnology is the key to ending world hunger. In the chapter on developing countries and that on the European Union, however, we have also seen the way in which pressure for regulation and restriction has grown. The World Bank has, in particular, come under intense pressure from concerned NGOs to end its endorsement of agricultural biotech. As a spokeswoman for one such organisation, Marcia Ishii-Eiteman, staff scientist at the Pesticide Action Network North America (PANNA), asserts, the bank should 'immediately withdraw its support for ongoing and planned GE projects and redirect those funds to programs that are proven, low-cost, and ecologically-based, such as soil fertility and Integrated Pest Management'.[1]

In view of this tension and the difficulties of resolving essentially irreconcilable positions, it is not surprising that the most comprehensive and definite instrument yet applied to biotechnology on the international level is that governing trade in ' living modified organisms', those GMOs as yet unprocessed into foodstuffs. While developing countries are divided over the potential usefulness of GM foodstuffs in tackling food supply problems, they are united in their desire to control what does and does not enter their territory in a form which could affect not only their agricultural products but their entire ecology. Nor, on the other hand, is it surprising that this instrument, known for short as the Cartagena Protocol on Biosafety, has weaknesses as well as strengths.

## THE (CARTAGENA) BIOSAFETY PROTOCOL TO THE UNITED NATIONS CONVENTION ON BIOLOGICAL DIVERSITY

The UN Environmental Programme (UNEP) first committed itself to drawing up guidelines for the safe application of biotechnology in June, 1992, at the Rio Conference where Agenda 21, a comprehensive international agreement on environmental protection, was adopted. Chapter 16 of Agenda 21 sets the cautiously optimistic tone in which the subject tends to be approached in such fora, 'recognising', in UNEP's own words, that 'although biotechnology cannot provide solutions to all the fundamental problems of environment and development, it could nevertheless contribute substantially to sustainable development by improvements in food production and feed supply, health care and environmental protection'. Despite these advantages, modern biotechnology might well present dangers, while 'the community at large can only benefit maximally from biotechnology if it is developed and applied judiciously'. This meant that efforts must be made to 'ensure safety in biotechnology development, application, exchange and transfer through international agreement on principles to be applied on risk management and assessment'.[2]

The Cartagena Protocol, adopted seven years later, was clearly built upon the same philosophical and administrative foundations as can be detected beneath these statements. The Convention on Biological Diversity (CBD) in 1995 established an Ad Hoc Working Group on biosafety which would eventually lead to the Protocol. The CBD's motivation in setting up this group was the widespread feeling that it was high time UNEP finalised its International Technical Guidelines for Safety in Biotechnology. It was hoped that these might provide interim guidance until a Protocol could be agreed. In fact, the Protocol when it finally arrived does follow the UNEP Guidelines in many of its details, and the Guidelines can be seen as the seedbed of the Protocol, a set of principles which could be judged on their record and which gave governments more confidence in adopting a more binding set of rules.[3]

The Cartagena Protocol came into force in September, 2003, following its ratification 90 days earlier by the tiny state of Palau. The Protocol stipulated that it would enter into force only when 50 states had ratified. The honour of being number 50 could have gone to the European Union, but last-minute disagreements between the European Parliament and the Commission delayed the ratification slightly and handed the historic opportunity to Palau.

The Protocol governs trade in ' Living Modified Organisms' (LMOs) intended for use as food and feed. LMOs would include seeds for planting, animals and fish, micro-organisms, and vegetables for processing, but not finished products. Thus, soya beans in their raw state are covered; processed and added to, say, tins of stew or frozen pies, they are not. Food safety issues are governed by other international agreements which, unlike Cartagena, were not introduced specifically to deal with GMOs but which have, in some cases, been adapted to do so. LMOs are not covered whilst in transition across the territory of other countries to a Party of import; or if they are destined for fully contained use, and will thus not in principle enter the broader environment. In addition, LMOs intended for use as pharmaceuticals for humans are excluded by specific provision on the grounds that they are already adequately covered by general regulations covering trade in medicines.[4]

Article 1 of the Protocol states that its objective

> is to contribute to ensuring an adequate level of protection in the field of safe transfer, handling and use of living modified organisms resulting from modern biotechnology that may have adverse effects

> on the conservation and sustainable use of biological diversity, taking also into account risks to human health, and specifically focussing on transboundary risks.[5]

This is the familiar style of international negotiation, the clear product of a process which has begun with differences which may have seemed irreconcilable but which, in the end, proved to be no such thing. Such compromises are achieved when the imperatives pushing the process prove more powerful than the initial lack of agreement, producing more and more concessions from both sides. In this case, the great majority of developing countries were concerned to retain control of potentially problematic products entering their territory whilst not excluding themselves from possible benefits. In other words, they had not made up their minds as to whether GMOs might offer them advantages or not. They recognised that not enough was as yet known for any definitive judgement to be passed and were determined to sit on the fence before deciding themselves which way to jump, rather than being pushed one way or the other by the warring forces of the North and West. For their part, GMO-producing countries, and in particular the developed countries which controlled or sought to control the production and distribution of GM seeds and products, were anxious not to alienate potential trade partners. The glaring exception was the United States, by far the world's biggest exporter of GM products which, however, claimed to see no reason to subject these novel products to any specific regime, denying not only their potential hazards but their very novelty itself.

Unsurprisingly, the US was not among the 103 countries which had signed the Protocol by the time of its implementation, nor had it any intention of either signing or ratifying the instrument. According to a detailed analysis of the Protocol by international jurist Jeffrey Waincymer, the United States might well be reluctant to sign even if it did not adopt such an aggressive and uncompromising attitude to trade in GMOs. A principle of international law embodied in Article 30.4 of the Vienna Convention – a sort of treaty on treaties – is that 'when parties to a later treaty do not include all of the parties to the earlier one, then as between the parties to both, the earlier treaty applies only to the extent that its provisions are compatible with those of the later treaty'. However, where 'two parties are Members of the WTO Agreement and only one of those is a member of the Protocol, then the WTO Agreement prevails as between them'. Waincymer – who is less than sympathetic to the Protocol and a firm supporter of the WTO – points out that this order of precedence 'would provide a strong disincentive for a WTO member to sign the Protocol…' This 'order of precedence' was reinforced by the decision of the ministerial conference of the WTO in Doha, Qatar, in November 2001, to add a rider to the conference's final declaration which made the results of any future negotiations on compatibility between WTO rules and multilateral environmental agreements (MEAs) such as Cartagena binding only on those countries which have signed the MEA in question.[6] The dissuasive effect this is likely to have on any country hesitating as to whether to sign the Protocol is compounded by uncertainty as to how the agreement will work in practice, because its provisions 'are certainly sufficiently unclear to make this a strong policy consideration'. Any room for manoeuvre is further restricted by the fact that the Protocol expressly forbids any 'reservations': countries must agree to every word of the text, or stay out. It would not be possible, therefore, for a country which wished to join the Protocol on condition that WTO rights and obligations always take precedence to have this stated as a condition of entry in any way which would be binding under international law.[7]

Article 1's compromising birth can be seen first of all in the fact that it is restricted to 'living' organisms, those which provide the most obvious threat to the environmental wellbeing of the importing country, a matter which falls squarely and relatively uncontroversially within the traditional ambit of national sovereignty. The latter part of this Article also smacks of compromise: in this case, between those who wished to restrict the Protocol – which is, after all, appended to a Treaty on Biodiversity – to strictly relevant environmental considerations, and those who did not want to miss the opportunity of introducing more comprehensive protective rules. The solution was the appending of other considerations which would 'also' be 'tak[en]… into account'. They are there, but accorded a status rather lower than the 'conservation and sustainable use of biological diversity'.[8]

This formula is repeated several times in the Protocol, as in Article 2, paragraph 2 of which enjoins 'the Parties' to 'ensure that the development, handling, transport, use, transfer and release of any living modified organisms are undertaken in a manner that prevents or reduces the risks to biological diversity, taking also into account risks to human health'.[9]

Article 2 also contains the important provision that states may go further than the Protocol stipulates: in other words, its provisions are to be regarded as minima. However, in allowing this the Protocol moves from compromise to something dangerously close to genetically modified fudge, for any such actions must be 'in accordance with that Party's other obligations under international law'.[10] The problem with this is that the Cartagena Protocol is clearly either in contradiction to the rules of the WTO, or a limited exception to them. If, as one would hope, international law and national administrations recognise that Cartagena has established an exception to liberalised trade, then any attempt to see it merely as 'minima' could prove an inadequate basis for defence of trade restrictions before the WTO. The Protocol does not make it clear which international instrument would, in the case of such conflicts, take precedence, notwithstanding its so-called 'savings clause', which states that the Protocol does not affect obligations under other international instruments. In reality this does not mean that the WTO should always take precedence over the Protocol, because the WTO itself recognises, at least on paper, environmental concerns as legitimate reasons for restraint of trade. Not only that, but whereas the text of the Protocol clearly states that it 'shall not be interpreted as implying a change in the rights and obligations of a Party under any existing international agreement', this is swiftly followed by the seemingly contradictory assertion that this 'is not intended to subordinate the Protocol to other international agreements'.[11] The crucial question boils down to this: who would have the right to resolve any dispute between the two instruments? According to the International Institute for Sustainable Development (IISD),

> While the WTO's Committee on Trade and Environment has expressed its preference for disputes arising from an MEA (Multilateral Environmental Agreement, of which Cartagena and its parent Convention are examples) to be settled within the MEA, serious trade-related disputes under this Protocol would almost certainly end up in the WTO.[12]

If the rules allegedly violated are those of the WTO, it is hard to see how anyone other than the WTO can investigate the allegations, pass judgements and order violations redressed or punishments inflicted. The only way to avoid this would be a clear statement that MEA rules took precedence, or that they created recognised exceptions to normal WTO rules. We have no such declaration. What we have been left with are two contradictory sets of obligations, leaving, in effect, the WTO to decide each case on its merits, according to the principle that genuine environmental measures are allowed, but using them purposefully to restrain trade is not.

This does not mean that WTO rules will always be used to ride roughshod over the obligations and rights contained within the Cartagena text. The Protocol clearly states that it is 'not subordinate' to other agreements. This is unequivocal, and although it is balanced by the seemingly contradictory 'savings clause', it is too clear-cut for the WTO to be in a position to ignore it. Optimists argue that what will happen in practice is that the Protocol will become 'the framework within which LMO trade will be judged in the WTO'. This would almost certainly be the case were it not for the refusal of the United States to sign up. As things stand, however, those who have criticised the Protocol's lack of clarity as to which set of obligations should take precedence have a point. Lawyers particularly dislike the Preamble, which, on the one hand, states that the Protocol 'shall not be interpreted as implying a change in the rights and obligations of a Party under any existing international agreements', and, on the other, follows this immediately by stating that this 'is not intended to subordinate this Protocol to other international agreements', a statement which one legal analysis complains 'arguably negates' it.[13]

This may seem to be nothing more than a rather short-sighted attempt to put off conflicts which will eventually need to be resolved. This impression is, indeed, compounded by the fact that the Protocol specifically leaves unanswered the difficult but essential questions of, on the one hand, liability and redress in the event of damage caused by LMOs and, on the other, of what to do if a party fails to comply with requirements.[14] Previous experience told the negotiators that any attempt to work out a liability system would simply leave the attempted Protocol mired in the mud of disagreements great, small, and nit-picking. They therefore established a four-year period from implementation to the establishment of a liability regime, meaning that it must be in place by 2007. On this point the IISD was pessimistic, concluding that 'a meaningful mechanism in the end is not a foregone conclusion'.[15] In some ways this seems the worst fudge of all, for without liability and redress it is tempting to look at the (sometimes) fine words of the text and see only empty rhetoric. Unfortunately, however, if recent events have demonstrated anything, they have shown once again that international law works only through the goodwill of those who agree to obey it, even if this is a goodwill based less on altruism than on a sort of social contract:

we obey this law now, even though it is onerous, because tomorrow we may wish another to obey a law which is onerous to them, and we want to live with the order and security of a law-based system, even though it may not in every instance favour our immediate interests.

The core of the problem dealt with in the Protocol is that, as Waincymer puts it, 'fair and efficient trade related environmental measures need to properly consider and balance aspirations towards environmental protection with the need to ensure that such measures are not utilised for trade protectionist purposes...'[16] The Protocol can fairly be accused of leaving this issue to be resolved at a later date, of fudging. In this case, however, there is a defence of this confection: GM fudge may in itself be less than nutritious, but at the very least it may keep the patient alive until something more substantial can be found for her to eat. This is not the use to which fudge is generally put, but in this case it may be defended as something more than a despairing attempt at delay, for information on GMOs is accumulating rapidly. If those on the two broad sides of the debate honestly believe their arguments, then they will welcome, or at least tolerate such a delay as allowing more evidence to be gathered for prosecution or defence. The spirit of compromise motivating the Protocol can therefore be attributed not only to the need to resolve wide differences of opinion, but also to the widespread feeling that the issue will become easier to resolve when more is known.

The fact that any agreement at all was reached came as a complete surprise to most of the people involved, especially those from the environmental movement. Negotiations had begun in 1999 in Cartagena de Indias, Colombia, but were not completed until the following year when the Parties to the UN Convention on Biodiversity reassembled in Montreal. It was during this period that five broad coalitions of interests emerged: the 'Miami Group' consisted of the US (which, however, did not officially participate in the negotiations, holding only 'observer status'),[17] Canada, Australia and three Latin American countries: Chile, Argentina and Uruguay. These countries followed the US line of insisting on the primacy of WTO rules and their application to GM organisms and products. Any specific regulation should be limited to cases where the organism in question posed a proven risk to biodiversity. They opposed the incorporation of the precautionary principle and any reference to 'socio-economic' factors. In the extreme opposite camp was the 'Like-minded Group' of developing countries. They wanted every country to have the right to refuse imports of GMOs and all products thereof. In order to achieve this, moreover, they wanted assistance for countries which – generally through poverty and underdevelopment – lacked the capacity to take such decisions in an informed way and police them once taken. It was these countries which demanded effective risk assessment and management provisions, liability and redress mechanisms, the explicit inclusion of the precautionary principle and of a reference to socio-economic considerations.[18]

Two further blocs stood between these two extremes: the European Union did not back the Like-minded Group on every point, but agreed that the precautionary principle should be incorporated and that no savings clause should be included which gave automatic priority to WTO rules. Underlying this position was the European Commission's increasing – if reluctant – acceptance that the Council and Parliament were moving towards the adoption of the world's strictest system of regulation, and its consequent determination not to allow this to undermine the Union's competitiveness. If most of the rest of the world showed itself willing to accept a similar degree of stringency, this would be to the EU's commercial advantage. The US, being the world's major exporter of GMOs and GM products, wanted a system which would be as near to a free-for-all as politically feasible. The real division emerging as talks progressed was clearly much more to do with actual status in relation to the products of agricultural biotechnology than it was with level of development or any ideological consideration: importers were closer to the EU, exporters to the US, and this gave the misleading impression that the EU was more inclined to solidarity than its rapacious rival across the Atlantic.[19]

The second bloc to emerge between the two extremes became known as the 'Compromise Group' and included developed countries (Japan, Norway and Switzerland); the NICs, South Korea and Singapore; and a developing country, Mexico. Eventually, this bloc also won the support of New Zealand. The Compromise Group wanted neither a free-for-all nor a system which would hinder trade in GMOs beyond what was strictly necessary to protect biodiversity, though its members naturally differed as to how such necessity might be defined. This group has been described as playing a role which 'was to prove critical in the final discussions'.[20]

One major sticking point concerned just what organisms and products should be included. Should all 'LMOs' be included? What about products? If all GMOs and GM products were included, should procedures be equally stringent in all cases? Wouldn't this have a detrimental effect on international trade? How would the rules be enforced and who should foot the bill if LMOs really did cause damage?[21]

In the end, despite wide and sometimes bitter differences of opinion, a compromise was found which only the United States found completely unacceptable. While this weakens the Protocol's impact, it became increasingly clear during negotiations that any agreement to which the US would be prepared to accede would be worthless. Moreover, neither the Clinton Administration nor any likely successor would be likely to change this position, and even if they did, Congress would never approve ratification. Though the agreement was initially greeted with a chorus of approval from almost all sides,[22] it quickly came to be seen as a victory for the more cautious, more environmentalist approach and, by those who favoured a complete ban on GMOs whenever and wherever this could be achieved, as a step forward. Whatever its limitations, then, the Protocol can be regarded as an important step in internationalising resistance not, perhaps, to GMOs as such, but to the United States' government and US corporations' insistence that they present no problems whatsoever and are 'substantially' the same as their non-modified equivalents.[23]

At the centre of the system of controls established by the Cartagena Protocol stands the so-called Advance Informed Agreement (AIA) procedure. This requires that, 'prior to the first intentional transboundary movement of living modified organisms for intentional introduction into the environment of the Party of import', a number of steps be taken. Firstly, 'the Party of export' – the country under whose jurisdiction the exporter operates – must inform, in writing, the 'competent national authority of the Party of import' that the movement is about to take place.[24] The information which must at this stage be provided is listed in Annex 1 of the Protocol, and includes details of the exporter and importer, a detailed description of the LMO, the technique used to produce it and of the resulting modification, suggestions for safe handling, and the regulatory status of the LMO within the Party of export's territory.[25] For commercial reasons, the Party of export may require that some of the information supplied be treated, for commercial reasons, as confidential.[26] Secondly, the Party of import must give a reply within 90 days as to whether the transboundary movement may proceed and, if the answer is yes, whether any conditions are attached. The reply may, however, constitute nothing more than a notice that the

Party of export must wait for written consent, in which case the Party of import has 270 days to do one of five things. It may approve the import, unconditionally or subject to provisos, or it may refuse approval; and it may request additional information, and/or extend the deadline by a stated period.[27]

Importantly, the Party of import may base its decision on the precautionary principle, as

> lack of scientific certainty due to insufficient relevant scientific information and knowledge regarding the extent of the potential adverse effects of a living modified organism on the conservation and sustainable use of biological diversity in the Party of import, taking also into account risks to human health, shall not prevent that Party from taking a decision...[28]

Moreover, the very first line of Article 1 of the Protocol refers to 'the precautionary approach contained in Principle 15 of the Rio Declaration on Environment and Development'.[29] The corollary is that Parties of import 'may, at any time, in the light of new scientific information', change its decision. Parties of export may request that a party of import review a decision in the light of new scientific information or a 'change of circumstances...that may influence the outcome of the risk assessment upon which the decision was based'.[30]

The precautionary principle is already fully recognised at international level in relation to trade in food and other agricultural products and, in that respect, Cartagena may not seem to do any more than confirm its importance. The WTO's Agreement on Sanitary and Phytosanitary Measures (SPS) unequivocally incorporates the precautionary principle in obliging Parties to take into account, when deciding to allow any living organism into their territory 'the potential damage in terms of loss of production or sales in the event of the entry, establishment or spread of a pest or disease'.[31] However, the detailing of a risk assessment procedure, and the adding of risk management (and guidelines for such) to the mere assessment of risk, represent important improvements to the definition of the principle and how it should work. The precautionary principle has always run the risk of itself being nothing more than a permanent fudge: these aspects of the Cartagena Protocol move us further away from that danger. In addition, in explicitly allowing Parties to take socio-economic factors into account when taking their decisions, it adds a most important consideration to those on which the precautionary principle may be based, one which gets us mercifully away from the absurd and dishonest pretence that any and every judgement must be 'based on sound science'. Finally, the Protocol lays down definite procedures which must be followed in the light of new scientific evidence becoming available where a lack of such evidence has previously forced regulators to rely on the precautionary principle; and it obliges exporters in some cases to cover the cost of risk assessments. In these ways it can be seen as going beyond the SPS and even, just possibly, beyond fudge. The International Institute for Sustainable Development, which has subjected the Protocol to close examination as a piece of international law, concluded that the way in which it has incorporated the precautionary principle is significant because of 'provisions [which] fill in some of the gaps in the SPS Agreement. They enrich the SPS by adding details that help operationalize the precautionary principle in the context of LMOs...' Furthermore, it 'arguably establishes the precautionary principle as a principle of international environmental law and perhaps – since it can be used based on human health and socio-economic considerations – of customary international law'.[32]

The risk assessments on which decisions as to whether to allow the import of an LMO and what conditions to place on such import must follow rules laid out in a further annex to the Protocol: it must be conducted 'in a scientifically sound and transparent manner' and 'on a case-by-case basis'; risks posed by an LMO must be compared with those posed by any non-modified equivalent; novel characteristics having potentially harmful effects must be identified, and an evaluation conducted as to whether 'taking into account the level and kind of exposure' of the receiving environment, these effects are likely to be realised and how damaging any such realisation might be. On the basis of this risk assessment the competent authority of the Party of import makes 'a recommendation as to whether or not the risks are acceptable or manageable, including, where necessary, identification of strategies to manage these risks'.[33] The Protocol also provides for emergency measures in the event of unintentional transboundary movements[34] and lays down rules governing 'handling, transport, packaging and identification' of LMOs.[35] Importantly, though the emphasis throughout is, understandably, on 'biodiversity' (with the repeated rider that 'human health' may also be taken into account) a separate article allows Parties to 'take into account...socio-economic considerations arising from the impact of living modified organisms on the conservation and sustainable use of biological diversity, especially with regard to the value of biological diversity to indigenous and local communities'.[36]

The inclusion or exclusion of LMOs intended for direct use as food or feed, or for processing (dubbed 'LMO-FFPs' during negotiations) had been one of the major obstacles standing in the way of an agreement, and the formula finally arrived at was inevitably a compromise. On the one hand it was held that, as the Protocol could deal with no issues which did not touch upon biodiversity, commodities should not be included except where they are intended for introduction into the environment. Against this was the argument that, intentional or not, unprocessed organisms would end up in the environment and this would have implications not only for biodiversity as such, but for the impact of biodiversity (or its loss) on human health. The resolution was that whilst LMO-FFPs would not be subject to the AIA procedure, a less rigorous version of the same procedure would be created specially for them. Instead of the Party of export being responsible for the notification of intent to move LMOs into the Party of import's territory, it is up to potential importers to develop and make known their regulatory requirements. Parties of export are not required to wait for a response from the Party of import before proceeding: provided they know what a Party's regulations require, they may proceed after a simple notification of intent. An important requirement of the Protocol is that exporters conduct studies, for the financing of which they are themselves responsible, in support of their applications; for exporters of LMO-FFPs, this is not the case. However, shipments of commodities which 'may contain' LMO-FFPs must be identified as such in documentation accompanying the consignment, a requirement which was accepted as an alternative to the mandatory segregation which the most cautious countries would have liked to have seen.[37]

The application of the AIA procedure to such a large volume of commodities was recognised as a difficulty, and a 'simplified procedure' was provided for in cases where the organism involved has already undergone safety assessments or has long been in use without causing problems. In addition, if the LMO is 'intended for direct use as food or feed, or for processing', developing countries or those with economies in transition may apply for financial and technical assistance in coming to a decision.[38] The organisation and provision of such assistance is one of the tasks of the Biosafety Clearing House (BCH), an international institution established by the Protocol in order to facilitate the information sharing on which the smooth working of the Cartagena system depends.[39] More broadly, the Protocol aims to facilitate capacity-building in such countries and to encourage the co-operation of richer countries in this.[40] Recognising that capacity building without attempting at the same time to raise public awareness of the importance of the issues involved could be a recipe for the creation of ineffective institutions, the Protocol also provides for measures to promote public participation in decision making.[41]

Developing countries have been encouraged by the Protocol to establish the regulatory and risk assessment framework it requires, but how effective these will be depends very much on the co-operation and goodwill of richer trading partners and remains to be seen. In the meantime, however, Article 18 does require Parties to insist on documentation which, in the case of LMOs intended for direct use as food, feed or for processing, declares that they 'may contain' LMOs 'not intended for intentional introduction into the environment'. This is an extremely important provision as it means that any shipment which could possibly contain traces of GM foods will need to be labelled, and that unless and until a party of import has the capacity to conduct effective risk assessment and management procedures, it will have the clear right, on the basis of the precautionary principle, to reject such a consignment.[42]

The coming into force of the Cartagena Protocol is only the beginning, of course. Difficult as the negotiating process proved, huge problems remain to be overcome, some of them inherent in international law, whose enforcement is never straightforward, and some of them deriving from weaknesses in the Protocol itself. The Protocol lacks its own mechanism for enforcement, leaving a huge area of uncertainty which must be filled in by juridical decision and precedent, a process which can leave the agreement either strengthened or dead in the water. No liability and compensation clause has as yet been agreed: whether it will be by the 2007 deadline remains to be seen. Precedents exist which might be developed: for example, the Basle Convention on the Control of Transboundary Movements of Hazardous Wastes and their Disposal contains structures for liability and redress which could provide a model, but these have themselves been criticised as insufficient.[43]

Though it remains to be seen how effective it will prove in practice, Cartagena nevertheless seems an impressive achievement. In that it came in the wake of Seattle, an event which drew the world's attention to the fact that a small group of powerful countries was attempting to create an international order where everything would be subordinated to the exigencies of the 'market', it can be seen as a triumph for activism. As the IISD concluded, Seattle's failure was in part caused by

> powerful concerted public protest against the elevation of commercial interests over socio-policy [sic] concerns, including the environment. Also responsible was a negotiating process that did not take serious account of the interests of most of the WTO's developing country members. So soon after Seattle, and in the glare of public attention generated by activist NGOs, key governments clearly had no desire to undermine progress on a treaty that so directly aimed to protect the environment and build capacity in developing countries...[44]

Cartagena represents a step towards effective control of trade in GMOs, and in that it is to be welcomed. Of at least as great significance, however, is the fact that the original demand for such a treaty came from those developing countries who, before Seattle, had been told that unless they played the game of trade according to rules set by a small group of privileged participants, they would not get to play at all.

## OTHER INTERNATIONAL AGREEMENTS TOUCHING ON TRADE IN GMOs

As the environmentalists Peter Newell and Ruth Mackenzie concluded their analysis of Cartagena, the Protocol

> will necessarily interact, both at the international level and in the context of national implementation, with a range of other international and regional instruments and arrangements. These include not only the WTO regime, but also ongoing work within the Codex Alimentarius on foods derived from biotechnology, other relevant FAO agreements such as the International Plant Protection Convention (IPPC), and relevant work within regional and economic organisations, as well as proposals for considering biotechnology and biosafety within other fora

for which they offer the example of an intergovernmental panel suggested by some members of the OECD.[45]

### Codex Alimentarius

The Codex Alimentarius is described by its parent body, the United Nations Food and Agriculture Organisation, as 'the seminal global reference point for consumers, food producers and processors, national food control agencies and the international food trade.' Although Codex Alimentarius standards are non-binding, many are adopted wholesale by the WTO or other regulatory authorities, and most national governments rely on them to ensure that their exports do not fall foul of hygiene and health rules when they arrive at their destination. According to the *Financial Times*, Codex standards are 'used as benchmarks in World Trade Organisation disputes to help determine whether a country's safety standards represent a disguised barrier to trade'.[46] The Codex Alimentarius Commission, its governing body, established in 2000 a task force charged with drawing up an agreement on the risk assessment of GMOs. The resulting 'Principles for the risk analysis of foods derived from biotechnology' were adopted in March 2002 as a recommendation to the Commission itself, which accepted them without substantial modification at a meeting in Rome in July 2003. Although they stopped short of agreeing to the EU's request for an international system guaranteeing full genetic traceability, the guidelines were broadly welcomed as a step in the right direction. Under the agreement, a case-by-case safety analysis of GMOs and products of GMOs intended as food – including micro-organisms and the food products in which they are used – must be conducted before any such food is marketed. The effects of these foods, both intended and unintended, must be studied with a view to identifying any new hazards arising from the modification, whether deriving from actual toxicity or allergenicity or simply from a change in composition, such as reduced or enhanced content of a particular vitamin or other micronutrient. Because the assessment process necessarily involves uncertainties, precaution must be exercised and, when a marketing authorisation is authorised, post-market monitoring, including means to ensure traceability, considered in any cases where certainty is less than absolute. This last point represents a compromise which successfully resolved what had been a major sticking point. Allergenicity testing methods and other means of determining the safety or otherwise of particular GM plant species are subject to detailed requirements, though other analytical methods, risk management tools, and systems of monitoring are left to the Parties to decide, the Codex document merely providing guidance. Finally, some thorny issues were put off by means of a statement that other aspects of traceability were under consideration, and that the Codex committee on food labelling was looking at various draft recommendations on the labelling of GM foods. Of course, no system of labelling of GM-derived foods is acceptable to the United States, but in the current climate this may not prove an insurmountable barrier. Having adopted a standard for GM food safety which reveals the glaring inadequacy of the FDA's marketing approvals system, the Codex Alimentarius Commission may not allow US opposition on its own – and certainly not the absurd claim by US agribusiness interests that labelling would 'decrease global food safety' – to prevent its rules embracing labelling.[47]

As the Brussels-based *Environment Daily* noted, the guidelines marked 'a significant step forward in the global management of genetically modified...foods' which 'vindicate[s], at least partially, the EU's insistence on introducing a system to enable tracing [sic] GM foods'.[48] Greenpeace called them 'an important step forward' and noted how the United States' system of marketing approval, under which the FDA requires no case-by-case pre-market safety assessment, fell well short of what the Codex guidelines required. Consumers' International welcomed the fact that the new guidelines 'provide detailed procedures for determining whether a GM food contains new toxins or allergens, is altered nutritionally, or exhibits unexpected effects' and that they 'endorse the use of "product tracing" as a tool of risk management'.[49]

Internationally, the agricultural biotechnology industry responded to Cartagena and the new Codex Alimentarius guidelines not by backing the Americans' stonewalling approach, but with the more intelligent strategy of attempting to influence the actual systems developed by governments to protect the environment and public health from possible problems associated with the release of GMOs. As the industry body Crop Life International stated on the release of their own *Reference Guide for Biosafety Frameworks Addressing the Release of Plant Living Modified Organisms*, they now sought, in their own words, 'to help governments around the world develop national science-based risk assessment and risk management measures for the intentional release of plants that have been improved using modern biotechnology.' This at least represented a step forward from the 'Risk? There is no risk!' approach favoured by US trade negotiators.[50]

### The International Treaty on Plant Genetic Resources for Food and Agriculture

The International Treaty on Plant Genetic Resources for Food and Agriculture (ITPGR) is an adaptation of the FAO's International Undertaking on Plant Genetic Resources, which dates to 1983. In

1992, United Nations Conference on Environment and Development's so-called Agenda 21 included a commitment to strengthen the Undertaking and adapt it to the newly adopted Convention on Biodiversity. Nine years of negotiation and revision resulted in the adoption of a revised Undertaking,[51] which was adopted by the FAO in November, 2001, becoming the International Treaty on Plant Genetic Resources for Food and Agriculture.

The ITPGR is an attempt to ensure equitable access to the benefits of genetic resources in the areas of food and agriculture. It uses a number of instruments and practices to achieve this, including promoting the conservation of plant genetic resources through national and international collections of seeds and plants and defending the rights of farmers to re-use seed and to maintain access to traditional seed varieties. As with the Cartagena Protocol, capacity building in developing countries is seen as central to the success of the agreement.

A total of 92 countries signed the Treaty, and 40 of these must ratify it before it comes into force, a number which, at the beginning of 2004, appeared within reach.[52]

## PROPOSED UNITED NATIONS BAN ON HUMAN CLONING

In October 2003, the Vatican, the United States, Costa Rica and 60 other countries sponsored a resolution to the General Assembly of the United Nations calling for a ban on all forms of human cloning. The General Assembly was called upon to adopt a plan which would have led to a convention on human cloning to be drawn up and adopted within two years, with an interim prohibition on the research, development or application of any technique involving human cloning of any kind. Though the resolution would have been non-binding, it would certainly have dealt a powerful blow to those seeking to develop new medical techniques involving genetic manipulation. The cloning of early embryos for research purposes provides the basic raw material of these techniques, and to outlaw it would effectively be to outlaw gene therapy. The broad consensus among scientists has been to recognise the ethical issues involved while rejecting the idea that a very early embryo is entitled to the same protection as a viable human being. The position of the governments backing the ban on the cloning of embryos was often opposed to that of the scientific establishment in their own country. Nowhere was this more true than the United States, whose scientists continued to play a leading role in the field despite their government's hostility and the restrictions under which they must work. The National Institutes of Health is itself one of the participants in an international body founded in January, 2003 with the British Medical Research Council and equivalents from Singapore, Australia, Canada, Sweden, Finland and Israel to develop a broad programme of collaboration and agreed guidelines for funding, technical standards and ethical principles.[53]

Opposed to a total ban and agreeing only that the cloning of entire human beings should be put beyond the pale of the law, Belgium, the UK, Japan, China and around 20 other countries sponsored a counter-resolution to that effect, further arguing that a total ban would represent an unacceptable breach of national sovereignty. In the end this position in a sense won the day, though very narrowly, with its supporters making it clear that they would neither ratify nor respect a ban on all forms of human cloning. Neither these countries nor those who sided with the US and Costa Rica saw their resolutions carried. Instead, an Iranian proposal, backed by the Organisation of the Islamic Conference, to put off any decision for a further two years was carried by the narrowest possible margin, a single vote dividing the 159 countries which took one or the other side.[54]

The stalemate satisfied no one. During 2001, groups of scientists began to organise internationally to oppose a total ban whilst demanding that reproductive cloning be made illegal. Immediately prior to the UN vote, these scientists issued statements opposing a total ban and calling for the immediate imposition of a total ban on reproductive cloning. Their arguments were couched in more scientific terms than those of the US/Costa Rica group, which tended to emphasise the 'repugnance' people felt in relation to human cloning. While not dismissing this, the scientists, from more than 60 national academies including Britain's Royal Society and those of China, France and – notwithstanding their government's position – the United States, preferred to emphasise that animal experiments had shown that cloning led to miscarriage, oversize foetuses, birth defects, post-natal death, and severe health problems in those clones which made it past all of these hazards. By 2002, 33 countries had banned reproductive cloning and many more were considering doing so, with most of those which had not underlining that this was for technical reasons and not because they were in favour of the technique. The call for a ban on reproductive cloning won the support of UN Secretary General Kofi Annan and would have been accepted at least as an interim measure by most countries. The US and other extremist countries, however, feared that to accept this position would mean that the much more sweeping ban they sought would lose impetus.[55] In taking this position, they were of course exposing the world to the danger that someone, somewhere would, quite legally, clone a living (if not for long), breathing (though probably with difficulty) creature resembling in some respects a human being.[56]

## UNIVERSAL DECLARATION ON THE HUMAN GENOME AND HUMAN RIGHTS[57]

Adopted unanimously at UNESCO's 29th General Conference session in Paris in November, 1997, the Universal Declaration on the Human Genome and Human Rights (UDHGHR) states unequivocally that 'Everyone has a right to respect for their dignity and for their rights regardless of their genetic characteristics' and that because of this it is 'imperative not to reduce individuals to their genetic characteristics and to respect their uniqueness and diversity'.[58] Further, it

- forbids making 'the human genome in its natural state' a source of 'financial gains'[59]
- states that 'Research, treatment or diagnosis affecting an individual's genome shall be undertaken only after rigorous and prior assessment of the potential risks and benefits' and that the person concerned must give his or her 'prior, free and informed consent' and may decline to know the results of any tests[60]
- means that persons who do not have the capacity to exercise such rights may in general not be subject to any procedure which does not offer direct health benefit, though 'Research which does not have an expected direct health benefit may...be undertaken by way of exception, with the utmost restraint, exposing the person only to minimal risk and minimal burden and if the research is intended to contribute to the health benefit of persons in the same age category or with the same condition...'[61]
- forbids 'discrimination based on genetic characteristics that is intended to infringe or has the effect of infringing human rights, fundamental freedoms and human dignity'[62]
- guarantees confidentiality of an individual's genetic information, the right to compensation in the event of damage sustained as the 'result of an intervention affecting his or her genome'[63]

- places restrictions on research, forbidding anything which is 'contrary to human dignity' including 'reproductive cloning of human beings'[64]
- states that 'Benefits from advances in biology, genetics and medicine, concerning the human genome, shall be made available to all' and that research 'shall seek to offer relief from suffering and improve the health of individuals and humankind as a whole'[65]
- enjoins states to 'respect and promote the practice of solidarity towards individuals, families and population groups who are particularly vulnerable to or affected by disease or disability of a genetic character'[66]
- calls on richer countries to aid others with capacity building.[67]

The UDHGHR is an impressive document which undoubtedly provides a model for international law in this area. It is not in itself binding, however, as is shown by the strenuous and so far unsuccessful efforts to have a ban on human reproductive cloning adopted by the UN General Assembly.

## ATTEMPTS TO CURB BIOLOGICAL WEAPONS

Talks on strengthening the 1975 convention to restrict biological weapons have so far failed to reach agreement on specific measures.[68] They broke down for the first time in 2001 in a condition described by the *Financial Times* as 'bitter recrimination'. As usual, the problem was the refusal of the United States to participate in any serious attempt to find an effective solution to the problems, central amongst which was the lack of any system to ensure compliance. Whilst the US demands the right to search any and every facility in other countries, it refuses to open its own to foreign inspectors. Citing the risk of 'compromising the secrets of US drug companies and the bio-defence establishment', and alleging that a number of countries had undeclared bio-weapons, the United States eventually demanded that negotiations on a comprehensive protocol cease in favour of focus on certain achievable goals. When the talks reconvened in November 2002, it was to discuss proposals which, though they were officially brought forward by the Hungarian chairman Tibor Toth, looked suspiciously like the suggestions made by the US prior to the breakdown of the previous year's talks. In place of an attempt to find an effective treaty, Toth proposed instead the convening of a meeting in 2004 'aimed at improving the international ability to investigate and mitigate the effects if bio-weapons have been used, and to improve international surveillance of disease outbreaks'. In addition, a meeting in 2005 'would cover scientists' code of conduct'. Not everyone was happy with this approach, but while many developing countries objected, arms control NGOs and experts urged them not to reject the talks completely. As Patricia Lewis, head of the UN Institute for Disarmament Research, said, 'In five years time the changes in biotechnology will be enormous so you've got to keep this subject on the international agenda.'[69]

The meeting did succeed in reaching a limited agreement, though it was one which almost entirely reflected the US approach. Three conferences would take place which would have the task of adopting 'specific measures to reinforce the treaty before the next review conference in 2006'. These would include 'criminalising breaches of the convention in national penal codes, reinforcing the security of pathogens, improving surveillance of and responses to disease outbreaks and adopting codes of conduct for scientists'.[70] Developing countries were grudging in their acceptance of the proposals. According to the *Financial Times*, they expressed their disappointment over their 'limited ambition' and, in a clear reference to the US, criticised the growing resort to 'unilaterally imposed solutions'. No further attempts to establish a system of inspection or other means to bring about compliance would be included, while other issues of interest to developing countries, principally technology transfer and scientific co-operation, were also shelved.[71]

## FURTHER READING

Morven A. MacLean et al., *A Conceptual Framework for Implementing Biosafety: Linking policy, capacity, and regulation* (International Service for National Agricultural Research (ISNAR) Briefing Paper 47, March 2002)

Ruth Mackenzie et al., *An Explanatory Guide to the Cartagena Protocol on Biosafety* (IUCN Environmental and Law Paper No. 46, 2003)

Sunshine Project *An Introduction to Biological Weapons, Their Prohibition, and the Relationship to Biosafety* (Backgrounder Series No. 10, April 2002)

Convention for the Protection of Human Rights and Dignity of the Human Being with regard to the Application of Biology and

Medicine: Convention on Human Rights and Biomedicine <http://conventions.coe.int/treaty/en/Treaties/Html/168.htm>

Universal Declaration on the Human Genome and Human Rights <www.unesco.org/ibc/uk/genome/projet/index.html>

# Conclusion

Biotechnology has confronted the world's legislators with a range of problems, some of them wholly new, others wearily familiar. In agriculture, the cultivation of GMOs threatens to reinforce inequitable structures of power and destructive inequalities of wealth, to make farmers still more dependent on corporate suppliers of agricultural chemicals and seeds, and to destroy attempts to develop sustainable and environmentally friendly systems of food production. No one has ever shown that GMOs are safe to eat, and there are numerous reasons to fear that they may not be, as a number of studies have pointed to possible adverse impacts of GMOs on human health.[1] An editorial in *The Lancet*, the journal of the British Medical Association, listed the possible problems as including 'allergenicity; gene transfer, especially of antibiotic-resistant genes, from GM foods to cells or bacteria in the gastrointestinal tract; and "outcrossing", or the movement of genes from GM plants into conventional crops, posing indirect threats to food safety and security'.[2]

## GMOs: IRREVERSIBLE CONTAMINATION OF THE WORLD'S FOOD SUPPLY

There are still more reasons to fear that GM crops may be leading to irreversible contamination of the world's food supply and of wild relatives. The biotech industry, agribusiness, and pharmaceutical corporations have used their power to prevent the necessary epidemiological studies and controlled trials from being conducted, and to suppress or denigrate any findings which they do not like. They have been less successful, however, when it comes to environmental problems. The British farm-scale trials of GM crops, concluded in October 2003, put an end to any idea that such problems were a product of the fantasies of fanatical eco-warriors. As well as posing a threat to the environment, GM plants' ability to contaminate neighbouring organisms could destroy the livelihoods of organic farmers and any whose sales depend on the GM-free status of their product.

The problems inherent in GM technology are compounded by the sheer complexity of the systems with which we are dealing, and the interdependence of their parts. Genetic engineers would like to be able to take a plant, tweak its genes, stick it in a field and watch it grow. But the field is no laboratory, it is part of a vast and ever-changing ecosystem. Alter one element and the consequences for the system as a whole may be anything from insignificant to devastating. One fear is that 'cascading trophic interaction' may occur, in which changes to a plant make it less (or more) suitable as food for a particular herbivore and this in turn affects the species which prey on the herbivore. Other species, such as parasites, may also be affected. The cascade may cause all manner of changes to an ecosystem, not all of which, probability and logic dictate, will be unnoticeable or benign. Ecological scientist Stephen Nottingham cites pollinating insects as a particular area of concern. 'If transgenic crops were to harm honeybees, for example via an insecticidal protein expressed in their pollen,' he warns, 'the pollination of a wide range of crops and native flowering species could be affected.'[3]

Each plant species and individual, whether cultivated or wild, forms part of a complex of ecological relationships linking the tiniest virus to the predator which tops the particular ecology's food chain. The results of even a small change are therefore extremely difficult to foresee and in many cases, short of some breakthrough in science, quite unknowable. The addition of a powerful and poorly understood tool which enables us to generate new species does, however, greatly compound this danger. Insect geneticist Marjorie Hoy summed up the current situation perfectly when she said that 'The science of putting genes in is far ahead of the risk assessment research.'[4]

The concentration of genetic engineering on herbicide resistance is no accident of science. The aim of the companies involved in such research is to make farmers doubly dependent on them. Farmers buying GM seeds are generally bound by contract not to save seed for replanting, and any plants resulting from the spread of GM seed to a neighbouring farmer's land must be removed by that farmer if he or she is to avoid being sued for patent infringement. At the same time, by conferring resistance only to a specific herbicide, a company can ensure that both seed and herbicide are supplied by themselves and no one else.

In 1996, throughout the world, only 1.7 million hectares of GM crops were grown.[5] By the end of 2002, industry sources were claiming that the total amount of land under cultivation globally for genetically modified crops had reached 58.6 million hectares, accounting for more than 20 per cent of soya, maize, cotton and rape. In total, there were 36.5 million hectares of soya, which accounted for a great majority of the world's transgenics. The figure for maize was 12.4 million hectares, with 6.8 million given over to cotton and 3 million to canola, with relatively small amounts of squash and papaya taking up the rest. No other GM crop plants were licensed for commercial use anywhere in the world, though many other plants were grown experimentally.[6]

Continued growth in both commercial and experimental cultivation was strong, despite a moratorium on new crop approvals in the European Union, the continuing reluctance of Brazil to countenance the technology, and widespread suspicion in Africa, the Indian subcontinent and the Far East. Soya was of particular importance, as around two-thirds of processed foods on sale in developed countries contain ingredients derived from it, a world total of around 110 million tonnes of soya beans being produced per annum. At the same time, the global market for GM crops was estimated by the industry body ISAAA – the International Service for the Acquisition of Agro-Biotech Applications – to have grown to around $4.25 billion. Soya is also one of the United States' principal exports to the European Union, accounting for a quarter of total agricultural products moving east across the Atlantic between the world's two biggest markets.[7]

Such trade, and the production and distribution of soya, maize, rice, wheat and the other major staples of the human diet, as well as of the various chemical inputs which intensive agriculture requires, are subject to increasing control by a small group of very large transnational corporations (TNCs). For example, by 2002, ten pesticide companies, based in the US and Europe, controlled 84 per cent of the $30 billion annual pesticide market. Five of these companies (Du Pont, Syngenta, Bayer, Monsanto and Dow) controlled 25 per cent of the global seed market and 71 per cent of all patents on agricultural biotechnology.[8]

Why have GMOs spread so far so fast, but only in the US and a small number of other countries? A friendly legislative environment is a *sine qua non*, and, in the United States, successive Administrations have used legislative action to smooth the way. Given the right conditions, however, farmers may well be genuinely attracted to transgenic crops for their labour-saving and claimed input-reducing qualities. In some cases they are misled by simple falsehoods, as when Monsanto claimed that their Bollgard cotton was not susceptible to bollworm damage.[9] In others, GM products have genuinely increased yields, or profit margins, at least for a while. Often, they appeared to present solutions to persistent and intractable problems. Yet just as often, these solutions have proved either illusory or, at best, short-lived. One United States Department of Agriculture survey concluded that, up to 2000, available GM varieties had not increased 'the yield potential of any variety'.[10]

Farmers surely understand their own business and are likely quite quickly to see through short-term yield improvements, pesticides which cease to function, or pests which disappear only to return or be replaced by others equally damaging. In some cases, however, by the time these problems become apparent farmers may already be trapped into contracts difficult or impossible to escape. The massive power of the corporations profiting from this enables them to get away, quite simply, with lying about the efficacy of the goods they are selling. This power enables them not only to make farmers offers they can't refuse, but also to manipulate the media and therefore public opinion, to buy the votes of elected representatives, and to undermine the independence of public officials, universities and learned journals.

## MEDICINE AND HEALTH CARE

The same advances in knowledge of genetics and molecular biology which have provided the scientific basis for agricultural biotechnology also have huge implications for medicine and health care. Yet here, too, there are serious doubts about the desirability and potential of an approach based on expensive technologies and poorly understood science. The Human Genome Project's discovery that human beings had between 30,000 and 40,000 genes has overturned the 'central dogma', the belief that one gene codes for one protein. Somehow, possibly as few as 30,000 genes code for around 750,000 proteins, a finding which pulled the rug out from under those who were seeking, whether for agricultural, medical or other purposes, to intervene in the genome of humans, animals, plants or micro-organisms. Far from one gene generating one protein to perform one function, a huge proportion of human genes are 'alternatively spliced', giving them the ability to generate not one protein but a range of them. Before knowledge of a gene can be put to medical use, therefore, the proteins for which it codes for, what those proteins do, the circumstances in which it resplices, and what might influence the various phases of its expression must all be studied. This science – dubbed 'proteomics' – is complicated stuff, its relationship with genomics neatly expressed in the title of the 2001 conference 'Human Proteome Project: "Genes Were Easy"'.[11] Yet it surely needs to be developed before medical biotechnology can move beyond its present hit-and-miss methodology.

Unfortunately for those who see gene therapy as the future of medicine, with rare exceptions it appears to be promising jam tomorrow, while the feeling is widespread that more ought to be being invested in the bread-and-butter medicine of today. While more than 500 clinical gene therapy trials are either ongoing or have been completed, only one gene therapy has as yet been given marketing approval, and that only in the People's Republic of China. Even enthusiasts admit that the 'available technology' is limited, or, as EuropaBio euphemistically confesses, 'not optimised yet in terms of efficacy and safety'. A 2001 report for the European Parliament noted that 'major technical problems remain to be overcome, most notably the lack of high efficiency gene transfer and the risk of adverse immune response provoked by using viruses as gene transfer vectors'.[12]

Much the same can be said of the development of therapies based on stem cells. Legislators must decide on the ethics of using stem cells taken from human embryos, while scientists attempt to find less controversial sources. Yet in the end the ethical difficulties will become significant only if the therapies based on stem cells actually work. Their track record is decidedly mixed. As the Nobel Prize winning British geneticist Sir Paul Nurse said recently, while stem cells may 'have exciting potential...their benefits have almost certainly been overstated, just like gene therapy ten years ago'.[13]

## Eugenics

The ghost of the eugenics movement and the Nazism and other repressive political ideologies which embraced it haunts most discussions of genetics in medicine. However well-intentioned it may seem, the possibility of interfering in the genetic make-up of individuals raises the spectre of Huxley's *Brave New World*. First comes the possibility of failure: what if we produced monsters? This is nothing, however, compared to the frightening implications of success.

The possibilities raised by genetic therapy pose the question of when 'cure' or 'preventative' becomes improvement or enhancement, when it steps over from the normal realm of medicine into what is generally understood by eugenics. The answer is not always easy and almost never clear-cut, which, in a world governed by the quest for profit and power, makes it open to manipulation. Therapies are being developed, for example, which would stimulate growth in children of abnormally short stature so that they would grow into adults of normal height. The obvious question this raises is 'What is normal height?' This is no trivial question: we can argue that it would be better if being short did not result in discrimination. Yet recognising that it does, or may bring psychological problems, can we also accept that parents who know that a son will never grow above 1.4 metres tall might feel inclined to submit their child to a therapy which was safe and effective and which would result in his growing to 1.7 metres?

But what if it were not always safe? Or what if it were expensive (and it will be)? Do we agree to socialise payment, or accept that in the future only poor people's children will grow up short? What if someone felt that, really, 1.7 metres wasn't enough, and that it would be a fine thing to take their child, of perfectly normal height, and ensure that he grew to be 1.9 metres? These questions would be interesting if we encountered them in an episode of *Star Trek*. They do not, however, apply to an imagined twenty-fourth century, but to the here and now, for this treatment is already available, and arguments as to who should get it are already raging.

Short of the horrors of eugenics are the more quotidian dangers of simply being ripped off. The tendency to define almost everything as a medical condition with a potential cure is already evident. Deviate from the standard model and you're ill. People who actually need a holiday, an improved diet, or some tangible reason to live are given instead a pill. If we can be convinced that there is a perfect human being, a sort of Platonic ideal to which we can and should aspire, then evil people could use this to bring about the sort of dystopia of which the Nazis dreamed. Or, what seems the more immediate danger, they can use it to sell us pills.

These pills are as likely to purport to push us towards the perfect psychological condition as they are to give us the perfect level of blood pressure or perfectly shaped lips. The rise of so-called 'behavioural genetics' in parallel with new discoveries in biology and the development and application of technologies based upon them has led to widespread fears that the ideas of the eugenics movement of the late nineteenth and early twentieth centuries are returning under a thin veneer of 'modern sophistication'.

### Cloning

There is no necessary link between human reproductive cloning and eugenics. However, the link exists in the public mind and for a very good reason: it is simply hard to see why, under current circumstances, anyone should interest themselves in cloning a human being unless the goal is 'genetic enhancement'. Assisted reproduction techniques have existed for a quarter of a century and improvement of them is ongoing. Cloning would be a hopelessly speculative and prohibitively expensive addition to these techniques. As the creators of Dolly the sheep, the world's first mammalian cloning success, have pointed out, 'genetic engineering can in principle change the nature of humanity, the meaning of *Homo sapiens*'.[14]

Most people are repelled by the idea of cloning, whether or not their grounds for rejecting it are rational. The exceptions tend to be attracted by the idea that it is possible to produce through cloning a perfect copy of themselves or a loved one. In fact, cloning does not result in perfect copies at all: environmental factors, beginning in the womb, ensure that no two human beings can ever be truly identical. The sole, if perverse, attraction of cloning turns out to be based on a myth.

New biotechnological techniques may raise complex legislative issues, but those relating to human reproductive cloning seem clear. The American social theorist Marcy Darnovsky surely speaks for a broad consensus when she concludes that the questions raised by the idea of human reproductive cloning not only 'add urgency to calls for strong legal bans on the production of cloned and genetically modified children' but also draw attention to the pressing need 'for effective social governance of other powerful genetic technologies'.[15]

### Diagnostics

The development of diagnostic biotechnologies also raises ethical problems which have increased in immediacy as techniques have become more available. Though fertility treatment is now in its third decade, for example, not until 2002 were couples in the UK with no history of hereditary pathology or disability offered the possibility of having embryos checked for genetic disorders prior to implantation. This was partly due to the technical difficulty and expense of the procedure, both of which have been reduced, partly because the number of conditions which can be detected at the embryonic stage is growing, so that the procedure has simply become more useful, and partly because of ethical considerations. Simon Fishel, director of the Centre for Assisted Reproduction in Nottingham, explaining the new possibilities, said that

> We have previously been able to test embryos to diagnose a genetic condition in the family. Now for the first time in this country we are able to screen embryos for a number of chromosomal disorders. The purpose is to achieve more pregnancies, fewer spontaneous abortions and fewer affected offspring.

Fishel has undoubtedly considered the ethical issues relevant to this and decided that his only real problem is that fertility treatment is a demanding business, especially for the prospective mother. His priority is understandably her wellbeing, and he believes this overrides other considerations, once it is accepted that women with difficulty conceiving (either personal, or because their partner has a fertility problem) have a right to be given all the assistance modern medicine can provide in their attempt to become pregnant. It would be surprising if a doctor whose specialism is assisted reproduction did not take such a view. Yet it is one which many other people would challenge.[16]

There are fears amongst disabled people that all disability will come to be treated as something to be eradicated. There is also concern that more and more 'conditions' will come to be defined as pathological, in the way that the pharmaceutical industry's desire to profit from Hormone Replacement Therapy led to a view of menopause as almost a disease, rather than a normal stage in a woman's life. Beyond these clear and present dangers, there is a fear that pre-implantation diagnosis coupled with either abortion or the development of curative pre-natal interventions will lead beyond the attempt to eradicate extreme disability (of the kind, for example, which leads to very early death) and ultimately to 'designer babies' and a new eugenics.

Ethical dilemmas are not unique to pre-implantation diagnostics. Any diagnosis which can be made before the appearance of symptoms – whether of an embryo, foetus, child or adult – is fraught with ethical dilemmas. In the words of molecular biologist Jean-Louis Mandel,

> because some of these diseases will start developing only at the age of 30 or 40...presymptomatic diagnosis is to say to a healthy person: 'You will develop this dreadful disease...in 10 or 20 years'. The international consensus rule is that this type of information should be given to the

> person at risk only if he or she...requests it, after having had enough time to discuss the issue involved with a multidisciplinary team, and to take a thoughtful decision.

Only about 0.1 per cent of the sequence determines those aspects of our differences which have a genetic element. As the *New Scientist* has warned, this means that 'accuracy is likely to be a tricky issue for personal genomics companies: get a single base pair wrong and a client may conclude they are about to die of a hereditary disease'.[17]

The problem of deciding just what the results of a genetic test should be taken to mean and how to respond to them is not confined to the lay public. In 2003, it was revealed that thousands of American women had been subjected to 'risky and unnecessary' tests during pregnancy following false diagnoses of the babies they were carrying as suffering from cystic fibrosis. Worse still, the false positives probably led some to have abortions which they would not have had had they not been misled. Researchers who uncovered these shocking occurrences believed that the blame must be shared between doctors who had misinterpreted test results and some of the companies responsible for the tests, companies which may have failed to follow clinical guidelines. Even if both the doctor and the testing laboratory get everything right, however, there remains a danger that the parents will overreact to a diagnosis which may sound more alarming than in reality it is.[18]

### Screening

Preventative medicine works on the level of whole societies as well as on that of the individual patient. Screening refers to the routine testing of entire populations, or vulnerable sections of populations, with the aim, in the main, of enabling early interventions where such would be helpful in saving lives or suffering. With advances in computer technology and the rapid development of what has been dubbed 'bioinformatics', the marriage of IT and molecular biology which has been described as 'the tool that lays bare the secrets within the code', screening seems set to become a cornerstone of twenty-first-century medicine. Public authorities in the US, the EU and Japan have already established the International Nucleotide Sequence Database (INSD) in an attempt to match gene to function. Yet there are those who doubt whether, in the end, it will all have been worthwhile, at least viewed from the collective, social standpoint of all human beings.[19]

**Pharmacogenetics**

The potential to know the precise genetic make-up of an individual points towards the possibility of adapting treatment, including drugs, to meet particular needs, whilst avoiding unwanted side-effects. Most drugs work for some individuals and not for others. Some patients may not respond at all to a medication which in other cases is clearly effective. Codeine has no effect whatsoever on certain people, because they lack the enzyme required to break it down into morphine in the brain. Some may respond, but only at the cost of unacceptable side-effects. In the case of a complex condition such as hypertension, where many factors may be involved, the usual approach is simply trial-and-error: the doctor will prescribe a drug, or combination of drugs, and then if the treatment does not work or unacceptable side-effects appear, change the prescription until he or she finds one which suits his or her patient. This can be unpleasant and even dangerous for the patient, whilst it is also, of course, wasteful and costly. If the right drug could be found first time, then these negative factors could be hugely reduced. Pharmaceutical corporations are attracted by the possibility that, if they could better predict who might suffer side-effects, they could win marketing approval for drugs which might otherwise be deemed too risky, as well as avoiding costly law suits and bad publicity. Adverse drug reactions (ADRs) are thought to cause up to 100,000 deaths every year in the United States alone.[20]

Small differences between the genetic make-up of individuals, known as single nucleotide polymorphisms (SNP) may, it is thought, hold the key to understanding differences in reactions to drugs. Some links have already been established, and it has become routine, in a few cases, to test for them before prescribing drugs. Others have their advocates for inclusion in the range of well-established SNPs and the drugs which should or should not be prescribed to people whose genetic makeup includes them. A broad division appears to exist between those of us who are 'fast' and 'slow' metabolisers. The latter may be more likely to suffer an adverse reaction to a whole range of drugs, persuading some that, even if attempting to identify every single relevant SNP in every single human being, or even 'groups at risk' would be a costly and non-cost-effective exercise, testing for this simple dichotomy would justify itself by significantly reducing the number of ADRs.[21]

It remains to be seen what pharmacogenomics really does have to offer. Certainly, reducing the lamentable number of ADRs is a laudable goal, but this might be better achieved by reducing the number of drugs prescribed in the first place by putting more emphasis on preventative medicine and training general practitioners to explore non-drug-based treatments (as well as educating patients to accept the validity of such treatments) before they reach for the prescription pad. The number of SNPs, which may be far greater even than current estimates indicate; the range of possible conditions; the number of available drugs; the fact that several SNPs may be involved in creating susceptibility to an adverse reaction; the range of possible adverse reactions, the various permutations of these figures produce astronomical totals which may make the technique quite unmanageable, or at least severely limit its utility. It is also difficult to see how clinical trials, without being increased in size to an impracticable degree, could identify or confirm possible SNP-related problems.[22]

## THE BOTTOM LINE

Genetic engineering is trumpeted by those who make money from it as a means to save the world from hunger and disease. It is in reality, however, a means to reinforce the imbalances of wealth and power which perpetuate malnutrition, hunger and ill health.

It is the private sector, pumped up with public money, our money, which is driving this technology forward. There will be no money in curing the poor or feeding the hungry. The huge profits needed to pay back the massive speculative investment of the last two decades will not be found in such projects, but in answering the demands of the market, which means, of course, of the at least relatively rich.

Multinational corporations in the field of agricultural biotechnology are seeking not only to enhance the profitability of existing systems already functioning as part of that market economy, but to draw into it the still sizeable minority of human beings which remains outside that system. The last few centuries have been characterised by the increasing appropriation into private hands of resources which were once available to all willing to labour to transform them into usable and valuable things. In much of the world this has already been achieved in relation to land, and water is now following land into the hands of big corporations and rich individuals. Agricultural biotechnology is the principal tool whereby this will be affected for the most fundamental of all factors of production, the self-replicating systems of life itself.

Legislators should reflect on the obvious truth that biotechnology's achievement will not be reflected in the well-fed, healthy faces of Bangladeshi children born into a world where hunger is something heard of only in Grandma's tales of the bad old days, but in the complacent, genetically rejuvenated features of overfed western consumers walking their cloned pooches across lawns of slow-growing grass.

That biotechnology developed from laboratory experiment to major industrial sector at the same time that the state and public authorities were abdicating more and more responsibilities in favour of 'market forces', when ideas of public service and social solidarity were widely seen as sentimental atavisms, is unfortunate. Never has the public good been in such need of protection from corporate irresponsibility. It should surely be the job of our elected representatives to create laws which provide such protection, or which at least enable us to protect ourselves. In some places, under pressure from a public increasingly aware of the problems associated with biotech, they have gone some of the way to doing so. In others, they have become nothing more than the builders of the road along which the juggernaut of corporate-controlled biotechnology is moving, crushing all that stands in its way.

## FURTHER READING

Lewis Wolpert *The Unnatural Nature of Science* (London: Faber and Faber, 1992).

Stephen Nottingham, *Genescapes: The ecology of genetic engineering* (London: Zed Books, 2002). See also my review of Nottingham's book at <www.spectrezine.org/reviews/Nottingham.htm>

Daniel J. Kevles *A History of Patenting Life in the United States with Comparative Attention to Europe and Canada* (European Group on Ethics in Science and New Technologies to the European Commission/Office for Official Publications of the European Communities, 2002)

James D. Watson with Adrian Berry, *DNA: The secret of life* (London: Heinemann, 2003).

Barry Commoner, 'Unravelling the DNA myth: the spurious foundation of genetic engineering', *Harper's Magazine*, February 2002, <www.findarticles.com/cf_0/m1111/1821_304/82743069/p1/article.jhtml>

G.J. Persley *New Genetics, Food and Agriculture: Scientific discoveries – societal dilemmas* (International Council for Science (ICSU), 2003), <www.icsu.org>

Martin Teitel and Kimberley A. Wilson *Changing the Nature of Nature: What you need to know about genetically engineered food* (London: VISION paperbacks, 2000)

Jane Rissler and Margaret Mellon, *The Ecological Risks of Engineered Crops* (Boston, MA: MIT Press, 1996)

Academies of Science *Transgenic Plants and World Agriculture* (Report prepared under the auspices of the Brazilian, Chinese, Indian, Mexican, US and Third World Academies of Science, and the UK Royal Society, 2000), <www.royalsoc.ac.uk/files/statfiles/document-116.pdf>

Consumers' Association, Policy Report, *GM Dilemmas – Consumers and genetically modified food* (London: Consumers' Association, 2001)

Sarah Sexton *If Cloning is the Answer, What Was the Question? Power and decision-making in the geneticisation of health* (Sturminster Newton: The Cornerhouse, 2002)

Andrew Webster et al., *Human Genetics: An inventory of new and potential developments in human genetics and their possible uses: Final study* (Working Document for the Directorate General for Research of the European Parliament – Scientific and Technological Options Assessment (STOA), 2001)

Daniel Charles *Lords of the Harvest: Biotech, big money, and the future of food* (Cambridge, MA: Perseus Publishing, 2002)

Brian Tokar (ed.) *Redesigning Life? The worldwide challenge to genetic engineering* (London: Zed Books, 2001)

Mark Winston *Travels in the Genetically Modified Zone* (Cambridge, MA: Harvard University Press, 2002)

Peter Pringle *Food, Inc: The promises and perils of the biotech harvest* (London: Simon and Schuster, 2003)

Daniel Swartz and Helen Holder (eds) *Of Cabbages and Kings: A cartoon book on genetic engineering* (Amsterdam: A Seed Europe, 2nd edition, 2000)

Bill McKibben, *Enough: Staying human in an engineered age* (New York: Henry Holt, 2003)

Jon Beckwith *Making Genes, Making Waves: A social activist in science* (Cambridge, MA: Harvard University Press, 2002)

*OECD Biotechnology Update*, <www,oecd.org>

UK Human Genetics Commission, <www.hgc.gov.uk/>

UK Human Genetics Advisory Commission, <www.dti.gov.uk/hgac/>

Nuffield Council on Bioethics, <www.nuffield.org/nioethics>

US National Bioethics Advisory Commission, <http://bioethics.gov/cgi-bin/bioeth_counter.pl>

American Society of Human Genetics, <www.faseb.org/genetics/ashg/ashgmenu,htm>

Human Genome Project, <www.nhgri.nih.gov/>

Genetics Virtual Library, <www.ornl.gov/TechResources/Human_Genome/vl.html>

International Bioethics Committee – UNESCO, <www.unesco.org/ibc/>

International Association of Bioethics, <www.uclan.ac.uk/facs/ethics/iab.htm>

European Society of Human Genetics, <www.eshg.org/>

European Group on Ethics in Science and New Technologies to the European Commission, <http://europa.eu.int/comm/secretariat_general/sgc/ethics/en/gee_en.htm>

# Glossary of Terms and Abbreviations

## POLITICS AND LAW

| | |
|---|---|
| AATF | African Agriculture Technology Foundation |
| AIBA | All India Biotech Association |
| APHIS | Animal and Plant Health Inspection Service of the United States Department of Agriculture |
| ASEAN | Association of South East Asian Nations |
| BCH | Biosafety Clearing House |
| BEUC | Bureau Européen des Consommateurs: main EU consumer lobby |
| BIO | Biotechnology Industry Organization (US) |
| CABI | Collaborative Agriculture Biotechnology Initiative (US) |
| CBAC | Canadian Biotechnology Advisory Committee |
| CFI | Canadian Foundation for Innovation |
| CGIAR | Consultative Group on International Agricultural Research |
| CIAR | Canadian Institute for Advanced Research |
| Charter of Fundamental Rights (EU) | Appended to the Treaty on European Union, a potential 'Bill of Rights' but one whose legal status is uncertain. |
| Communication | Official statement of the European Commission sent as advice to the member states, the Council, and/or the European Parliament |
| CONABIA | Argentina Commission of Assessment in Biotechnology |
| CBD | Convention on Biological Diversity |
| COST | Committee on Science and Technology |
| Council of Europe | Not an EU body, but one which now unites almost all European countries. Largely concerned with human rights issues. |
| Council (of Ministers) | The body which directly represents the governments of the EU member states. Made up of the relevant minister from each |

| | |
|---|---|
| | country, depending upon the subject, so that one reads of, for example, 'the Environment Council', the 'Agriculture Council', and so on. |
| Crop Life International | International lobby for agricultural biotech industry |
| CRIs | Crown Research Institutes (NZ) |
| DBT | Department of Biotechnology (India) |
| Directive | A law, agreed at EU level, which must be transposed into the legislation of each member state |
| ECJ | European Court of Justice |
| EEA | European Environment Authority (EU advisory body) |
| EFSA | European Food Safety Authority |
| EPA | Environmental Protection Agency (US); also Environmental Protection Act (India) |
| EPC | European Patent Convention |
| EPO | European Patent Office |
| ERMA | Environmental Risk Management Authority (NZ) |
| EuropaBio (previously SAGB) | The main industry lobby in the EU |
| European Commission | The European Union body responsible for proposing new legislation and overseeing the application of existing legislation, as well as governing external trade. Members appointed by member state governments. |
| European Community | Together with other entities to which the member states of the European Union automatically belong, the European Community forms the European Union |
| European Council | Composed of the heads of state and/or government of the member states of the European Union |
| EUP | Experimental Use Permit (US) |
| European Parliament | Elected by universal suffrage of the citizens of the European Union member states, the EP does not have all of the powers of a legislature. Its powers have been increased and broadened by successive treaties, however. |

| | |
|---|---|
| FAO | Food and Agriculture Organisation (UN) |
| FDA | Food and Drug Administration (US) |
| FIFRA | Federal Insecticide, Fungicide and Rodenticide Act (US) |
| FONSI | Finding of no significant impact (US term) |
| FSANZ | Food Standards Australia New Zealand |
| FSIS | Food Safety and Inspection Service (US) |
| GEAC | Genetic Engineering Approval Committee (India) |
| GRAS | 'Generally regarded as safe' (US term) |
| GTIT | Gene Technology Information Trust (NZ) |
| GTTAC | Gene Technology Technical Advisory Committee (Australia) |
| Habitats and Wild Birds Directives | The most important EU wildlife protection measures. |
| HAN | Highly Advanced National project (S. Korea) |
| HSNO | Hazardous Substances and New Organisms Act (NZ) |
| IBS | Intermediary Biotechnology Service |
| IDS | Institute of Development Studies (UK) |
| IMF | International Monetary Fund |
| IPPC | International Plant Protection Convention |
| IPRB | International Program on Rice Biotechnology |
| IOGTR, later OGTR | (Interim) Office of the Gene Technology Regulator (Australia) |
| IP | Identity Preserved |
| IPR | Intellectual Property Rights |
| ISNAR | International Service for National Agricultural Research |
| ITPGR | International Treaty on Plant Genetic Resources |
| IUPV | International Union for the Protection of Plant Varieties |
| JRC | Joint Research Council (of the European Commission) |
| LMO | Living Modified Organism |
| LMO-FFP | Living Modified Organism for direct use as food or feed, or for processing |
| MAFF | Ministry of Agriculture, Forestry and Fisheries (Japan) |

| | |
|---|---|
| MLHW | Ministry of Health, Labour and Welfare (Japan) |
| NARS | national agricultural research systems |
| NBS | National Biotechnology Strategy (Canada) |
| NERICA | New Rice for Africa |
| NLRD | Notifiable Low Risk Dealing (Australia) |
| NRA | National Registration Authority for Agricultural and Veterinary Chemicals (Australia) |
| OECD | Organisation for Economic Co-operation and Development |
| OGESA | Office of Genetic Engineering Safety Administration (China) |
| PANNA | Pesticide Action Network North America |
| Peace Corps | Official US body sending voluntary workers to developing countries |
| PRC | People's Republic of China |
| PVR | Plant Variety Rights, aka Plant Breeders' Rights |
| RCGM | Review Committee on Genetic Manipulation (India) |
| Recommendation | Non-binding guidance from the EU to its member states |
| Regulation | An EU law which applies throughout the Union's territory without needing to be transposed into the separate laws of the member states |
| RCGM | Royal Commission on Genetic Modification (NZ) |
| SACRED | Sustainable Agriculture Centre for Research, Extension and Development in Africa (Kenya) |
| Scottish Parliament | Elected by universal suffrage to represent the Scottish people, it enjoys wide domestic powers but not the full sovereign power of a national parliament. |
| TGA | Therapeutic Goods Administration (Australia) |
| Toi te Taiao: the Bioethics Council | Advisory body to NZ government |
| Treaty on European Union | The founding document of the European Communities was the Treaty of Rome. This |

| | |
|---|---|
| | has been amended several times, most recently by the Treaties of Maastricht, Amsterdam and Nice. Maastricht was officially known as the Treaty on European Union, and as amended at Amsterdam and Nice this is now the basic law of the EU. |
| UDHGHR | Universal Declaration on the Human Genome and Human Rights |
| UNDP | United Nations Development Programme |
| UNEP | United Nations Environment Programme |
| USDA | United States Department of Agriculture |
| USAID | United States Agency for International Development |
| USPTO | United States Patent and Trademark Office |
| Welsh Assembly | Elected by universal suffrage to represent the people of Wales, but lacks many parliamentary powers. |

## SCIENCE AND TECHNOLOGY

| | |
|---|---|
| antigen | substance causing the production of antibodies by the immune system |
| BGH/BST/rBGH | bovine growth hormone/bovine somatropin/ recombinant BGH – interchangeable terms for same substance fed to cows to increase their milk yield |
| blastocyst | An early stage in the development of an organism, in which the embryo takes the form of a hollow sphere. |
| Bt | *Bacillus thuringiensis* A soil-dwelling organism which produces a substance toxic to insects. Each subgroup targets a different insect. Modifying plants to express the toxin protects them against a specific predator and is one of the major goals of current genetic modification. |
| DNA | deoxyribonucleic acid: complex, very large molecule from which the nucleus and certain other parts of the cell are constructed. DNA contains a chemical 'code' which instructs the cell to produce proteins. |

| | |
|---|---|
| exon/intron | Sequences of DNA which (respectively) code for the production of proteins/appear to have no function ('Junk DNA') |
| ESC | Embryonic Stem Cell |
| gene | Unit of heredity transmitted from parent(s) to young, made up of sequences of DNA, influencing the appearance and behaviour of the organism. |
| HESC | human embryonic stem cell |
| introgression | Introduction of DNA from one organism to another |
| oocyte | An egg 'mother cell', that is, one which gives rise to an egg. |
| promoter | A sequence of DNA which encourages the introgression of a modified gene into the genome of the target organism |
| SNP | single nucleotide polymorphism |
| somatic nuclear cell transfer | Introduction of novel genetic material into the Hollowed out nucleus of an egg from which all genetic material has been removed. A basic technique of cloning. |
| totitpotent/ pluripotentmany | Able respectively to develop into any kind/ kinds of cell, generally a characteristic lost in an organism's very early development, though may be retained in some adult cells. |
| vector | A molecule used to introduce a modified gene into an organism. |

# Notes

## INTRODUCTION

1. *DNA: The story of life*, Channel 4 (UK), 8 March 2003.

## CHAPTER 1

1. European Commission *Life Sciences and Biotechnology: A strategic vision*, <http://europa.eu.int/comm/biotechnology/introduction_en.html>
2. Ibid.
3. See Steven P. McGiffen *The European Union: A critical guide* (London: Pluto Press, 2001), especially Chapters 2, 3 and 4 for an explanation of the legislative system and institutions of the EU.
4. A Directive is initially passed at EU level but must then be transposed into law by the member states according to their own legislative procedures; a Regulation becomes law throughout the Union as soon as it is enacted by the EU institutions, or on a date set by them, and need not be transposed.
5. For more details of the EU's legislative system, see McGiffen *The European Union*.
6. Directive 98/79/EC on in vitro diagnostic medical devices <www.ce-marking.org/directive-9879ec-IVD-MD.html>; and the *Directive* (not yet finalised and therefore unnumbered at time of writing) on setting standards of quality and safety for the donation, procurement, testing, processing, preservation, storage and distribution of human tissues and cells, <www.euractiv.com/cgi-bin/cgint.exe/615349–701?714&1015=9&1014=ld_humantissue>
7. *Charter on Fundamental Rights of the European Union*, <http://europa.eu.int/comm/justice_home/unit/charte/en/bibliography.html>
8. 'Biotechnology: Council calls for strategy to be applied soon', *Europe Environment*, 25 September 2003, pIII.4.
9. Figures quoted from Eurobarometer 55.2 *Europeans, Science and Technology*, 'Figure 1. Opinion of GMOs by EU country' in Sylvie Bonny, 'Factors in the Development of Opposition to GMOs', *IPTS Report* No. 73, April 2003, p.21.
10. 'Alarming risks of GM trialling', *Press and Journal*, 15 January 2003; 'Crop trials must stop, say doctors', *The Scotsman*, 19 November 2002; 'BMA report leads to call for halt on GM crop trials', *The Scotsman*, 20 November 2002; 'Doctors urge halt to GM crop trials', *FoEE Biotech Mailout*, Vol. 8, Issue 6, December 2002, p.3; the Scottish Parliament does not post such submissions on line. At time of writing, however, the whole of the BMA submission was available from Friends of the Earth Europe at <www.foeeurope/GMOs/index.htm>

11. Jonathon Carr-Brown, 'Conflict of interest over Sainsbury biotech cash', *Sunday Times*, 24 November 2002.
12. 'Belgian Environment Minister refuses two GMO field trials', Friends of the Earth Biotech Mailout, Vol. 8, Issue 3, June 2002, p.8; 'Belgium toughens line on gene crop trials', *Environment Daily*, 2 May 2002; 'Tighter rules urged for French GM crop trials', *Environment Daily*, 11 March 2002; Christian Babusiaux et al., *Rapport a la suite du debat sur les OGM et les essais au champ*, (Paris: Ministère de l'Ecologie et du Développement durable, 2002).
13. 'No agreement reached in German GMO debate', *Environment Daily* 1280, 5 September 2002.
14. Socialistische Partij *Dossier genetische manipulatie*, <www.sp.nl/dossiers/gentech/> (the author participated in the working group which wrote the Dutch Socialist Party's current policy statement on biotechnology); L.J. Brinkhorst, Minister for Agriculture, Environmental Management and Fisheries, on the *Report from the Temporary Committee on Biotechnology and Food Regarding the Public Debate 'Foodstuffs and Genes'* (Tweede kamer der Staten General, Vergaderjaar 2001–02, *Beleidsnota Biotechnologie Nr 9*) author's translation; 'Gentech ruimt het veld', *De Volkskrant*, 19 January 2002; the legal position in the Netherlands is described in English in a brief document from the OECD, *Regulatory Developments in Biotechnology in the Netherlands*, <www.oecd.org/ehs/netreg.htm>
15. Letter to members of the European Parliament from Helen Groome, Technical Adviser, General Union of Basque Farmers, 17 May 2002.
16. Sylvie Bonny, 'Factors in the development of opposition to GMOs and case-study evidence', IPTS, No. 73 (Seville; Joint Research Council, 2003); see also Sylvie Bonny, 'Why are most Europeans opposed to GMOs? Factors explaining rejection in France and Europe', *Electronic Journal of Biotechnology*, 5 (1), 15 April 2003, <www.ejbiotechnology.info>
17. Eurobarometer 52.1 *Europeans and Biotechnology*, <http://europa.eu.int/comm/research/pdf/eurobarmoeter-en.pdf;>
18. Pete Mitchell, 'UK government caught in GM dilemma', *Nature Biotechnology*, Vol. 21, No. 9, September 2003, p.957.
19. Paul Brown, 'Birds and bees: how wildlife suffered', *Guardian*, 17 October 2003.
20. Michael McCarthy, 'Proven: the environmental dangers that may halt GM revolution', *Independent*, 17 October 2003.
21. Brown, 'Birds and bees'.
22. 'The birds, the bees and biotechnology: Science triumphs as tests show GM crops threaten wildlife' *Financial Times*, 17 October 2003; see also 'Taking the long view', *New Scientist*, 25 October 2003, pp.46–9, an interview with Les Firbank, the scientist who headed the trials.
23. Council Directive 90/220/EEC of 23 April 1990 on the deliberate release into the environment of genetically modified organisms, <http://europa.eu.int/smartapi/cgi/sga_doc?smartapi!celexapi!prod!CELEXnumdoc&lg=EN&numdoc=31990L0220&model=guichett>; Council Directive 90/219/EEC of 23 April 1990 on the contained use of

genetically modified micro-organisms, <http://europa.eu.int/scadplus/leg/en/lvb/l21157.htm>

24. Directive 98/81 on the contained use of genetically modified micro-organisms, <http://europa.eu.int/smartapi/cgi/sga_doc?smartapi!celexapi!prod!CELEXnumdoc&lg=EN&numdoc=31998L0081&model=guichett>
25. Regulation 258/97 of 27 January 1997 concerning novel foods and novel food ingredients, <http://europa.eu.int/comm/food/fs/novel_food/nf_index_en.html>
26. Regulation 1139/98 concerning the compulsory labelling of certain foodstuffs produced from Genetically Modified Organisms, <http://europa.eu.int/scadplus/leg/en/lvb/l21233.htm>
27. Ibid.
28. Ed Dart, research director of Zeneca Seeds (described as 'Britain's biggest bioscience company') quoted in 'Laxer laws', *New Scientist*, 10 April 1993, p.11.
29. From a letter from K.T. Pike of ICI Seeds, 'The right balance', *New Scientist*, 4 July 1992, p.15.
30. Debora MacKenzie, 'Europe pushes for a genetically modified future', *New Scientist*, 11 June 1994, p.6.
31. Ibid.
32. Debora MacKenzie, 'Trouble in the wind over altered soya beans', *New Scientist*, 2 December 1995, p.12.
33. Bt maize has been genetically modified to express insecticidal toxins which are naturally present in the soil. By modifying a plant to produce its own, genetic engineers provide the organism with its own supply of protective toxin.
34. Antibiotic marker genes confer resistance to a selected antibiotic. They are not intended to play any role in the final product, but to accompany those introduced genes which are so intended, enabling scientists to spot whether the genes have been taken up by the genome of the target organism, and where. Promoter genes also accompany those genes believed to carry the traits which the genetic engineer is attempting to introduce. As their name suggests, these genes are designed to encourage the target organism to take up the genes with the useful qualities.
35. Andy Coghlan, 'Engineered maize sticks in Europe's throat', *New Scientist*, 6 July 1996, pp.7–8.
36. A brief overview of the legal situation on the eve of the substantial replacement of the legislation discussed above by the new legislative package which began with the adoption of Directive 2001/18 on the deliberate release of genetically modified organisms is provided by Marie-Martine Buckens and Lionel Changeur, *GMOs and Food Safety: Features from Europe environment* (Supplement to *Europe Environment*, Europe Information Service, October 2000); for details of releases, see 'Field releases in EU member states per year', <www.rki.de/GENTEC/GENENG/EU_YEAR_IE.HTM>
37. European Commission, *White Paper on Food Safety*, 1999, <http://europa.eu.int/comm/off/white/com99_719.htm>

38. Council and European Parliament Directive 2001/18/EC on the deliberate release into the environment of genetically modified organisms <http://europa.eu.int/eur-lex/pri/en/oj/dat/2001/l_106/l_10620010417en00010038.pdf>
39. European Commission, Memo/02/160 – REV. *Questions and Answers on the Regulation of GMOs in the EU*, 1 July 2003 <http://europa.eu.int/comm/dgs/health_consumer/library/press/press298_en.pdf>, p.1; Most law made at EU level no longer requires a unanimous vote of the member states, but is decided by a system known as Qualified Majority Voting (QMV) under which each member state is allotted a certain number of votes proportionate to its size, though smaller member states have proportionally more than the larger countries.
40. Vector: molecule used to transmit modified genes into an organism; may, for example, be a virus believed (or modified) to be harmless, leaving only its useful ability to invade and modify the cells of the target organism.
41. European Commission, *Questions and Answers*, pp.1–2; Jan-Peter Nap et al., 'The release of genetically modified organisms into the environment, Part 1: Overview of current status and regulations', *The Plant Journal*, Vol. 33, 2003, pp.1–18.
42. EuropaBio Press Release, 'New rules for biotech products enter into force', 17 October 2002; *European Commission Joint Research Centre Annual Report, 2002*, p.11.
43. European Commission, Memo/02/160 – REV. *Questions and Answers*, p.4.
44. *Environment Daily* 1705, 19 July 2004, <www.gmwatch.org/archive2asp?arcid=3612>
45. Regulation (EC) No 1829/2003 of the European Parliament and of the Council of 22 September 2003 on genetically modified food and feed, <http://europa.eu.int/eur-lex/pri/en/oj/dat/2003/l_268/l_26820031018en00010023.pdf>; Regulation (EC) No. 1830/2003 of the European Parliament and of the Council of 22 September 2003 concerning the traceability and labelling of genetically modified organisms and the traceability of food and feed products produced from genetically modified organisms and amending Directive 2001/18/EC, <http://europa.eu.int/eur-lex/pri/en/oj/dat/2003/l_268/l_26820031018en00240028.pdf>
46. 'Commission improves rules on labelling and tracing of GMOs in Europe to enable freedom of choice and ensure environment safety', European Commission Press Release, 25 July 2001, <http://europa.eu.int/comm/dgs/health_consumer/library/press/press172_en.pdf>
47. BEUC: The European Consumers' Organisation, Genetically Modified Food and Feed (labelling and traceability) 29 November 2001.
48. The EPP was the biggest of the European Parliament's political groups in the period 1999–2004. The EPP unites the EU's main centre-right current, including most conservative and Christian Democratic parties; the Parliament's other main group in this period was the PES (Party of European Socialists, centre-left). In addition, aside from a few very small groups, there were three medium-sized groups, the ELDR (liberals),

GUE-NGL (left and far left) and Greens-EFA (Greens and progressive nationalist/regionalist parties).

49. Julie Ruffle, 'GMO Report faces first plenary vote', *Rapporteur*, June 2002; Marie Woolf, 'Blair orders MEPs to block strict labelling of GM foods', *Independent*, 2 July 2002; British MEPs received a *Briefing Note on the Proposed Amendments to (the Trakatellis and Scheele Reports)* from the Department for Environment, Food and Rural Affairs (DEFRA), dated 29 May 2002, which gave detailed advice on how to vote on each amendment and could easily have been written by EuropaBio.
50. European Environmental Bureau, 'Memorandum: Progress report on GMOs', 30 January 2003, at <www.eeb.org/activities/agriculture/GMOs-Progress-Report-30–1–03.pdf>
51. This type of legislation can require up to three readings by the European Parliament. At the first, amendments require only a simple majority of MEPs present at the vote to be adopted; simultaneous to this First Reading, the Council, made up of representatives of the relevant ministers of member state governments, produces a 'Common Position'; the European Commission then responds to these. If no agreement between the three institutions can be found, the Parliament must hold a Second Reading where (a) only amendments which were successful at First Reading can be reintroduced, and (b) these must win an absolute majority of all MEPs, present or not, which at the time of these events meant 313 votes. If, after a Second Reading, the institutions are still at odds, a 'Conciliation' process is set in motion. If an agreement is then found, it is put to the full Parliament for a Third Reading when once again a simple majority is enough to secure adoption.
52. Christophe Schoune, 'Europe: Un accord au forceps sur les organismes genetiquement modifies', *Le Soir en Ligne*, 29 November 2002.
53. 'EU ministers agree on biofood labelling', Associated Press report, 28 November 2002; 'EU agriculture ministers vote on GM labelling' *Environment Daily*, 29 November 2002.
54. 'Observations du COPA et du COGECA de Réglement...concernant les denrées alimentaires et les aliments pour animaux génétiquement modifiés...et concernant la traçabilité et l'étiquetage des organismes génétiquement modifiés et...des produits destinés a l'alimentation humaine ou animale produits a partir d'organismes génétiquement modifiés...' Unsigned position paper dated 8 March 2002.
55. Letter from Beate Kettlitz of BEUC, the European Union consumer group umbrella body, sent to Euro-MPs and their advisers on 30 May 2002, listed BEUC's demands; see also Rob Edwards UK, 'GM-free food "is contaminated": Campaigners accuse US-based multinationals of holding the world to ransom in order to promote their products', *Sunday Herald*, 29 June 2003.
56. 'New GM thresholds in EU worry Canadian exporters', Reuters report, 2 December 2002.
57. Henry Miller and Gregory Conko, 'Brussels' bad science will cost the world dear', *Financial Times*, 14 August 2003; 'Zoellick: EU-US still a long way apart', *European Voice*, 11 September 2003, p.31.

58. Council Directive 98/95/EC of 14 December 1998 amending, in respect of the consolidation of the internal market, genetically modified plant varieties and plant genetic resources, <http://europa.eu.int/eur-lex/en/archive/1999/l_02519990201en.html>
59. European Commission *Adventitious Presence of GM Seeds in Seeds of Conventional Plant Varieties*. <www.projectgroepbiotechnologie.nl/toelating/download/EC-werkdoc-ggo.pdf>
60. David Byrne, European Commissioner for Health and Consumer Protection, 'Progress in building consumer protection', transcript of a speech given to the Committee on the Environment, Public Health and Consumer Policy, European Parliament, 2 October 2003, p.4, <www.europarl.eu.int/comparl/envi/pdf/speeches/20031002/byrne_en.pdf>
61. Commission of the European Communities Directorate General for Health and Consumer Affairs ('DG Sanco') Working Paper, *Adventitious Presence of GM Seeds in Seeds of Conventional Plant Varieties*, 2001; 'Report: Europe delays vote on biotech seeds rules', *St Louis Business Journal*, 27 October 2003, <www.bizjournals.com/stlouis/stories/2003/10/27/daily12.html>; no decision had been reached before the deadline for this book, though it may have been by the time you read it. If this is the case, you should be able to find the text at the general database of EU laws, at <http://europa.eu.int/documents/eur-lex/index_en.htm>; Jeremy Smith, 'EU approves GMO seed for planting across bloc', Reuters, 8 September 2004, <www.alertnet.org/thenews/newsdesk/L08433888.htm>
62. 'Biotechnology: Fischler stresses need to separate GMO and traditional crops', *Europe Environment*, 26 February 2002, Section 1, pp.7–8.
63. Commission of the European Communities, Commission Recommendation of 23 July 2003 on guidelines for the development of national strategies and best practices to ensure the co-existence of genetically modified crops with conventional and organic farming <http://europa.eu.int/comm/agriculture/publi/reports/coexistence2/guide_en.pdf>
64. 'Aventis GM maize: Scientist admits research was flawed' *Friends of the Earth Europe Biotech Mailout*, Vol. 8, Issue 3, June 2002 pp.2–5.
65. 'Genetic engineering: Austrian arguments on transgenic maize rejected', *Europe Environment*, 23 January 2001, Section iv, p.6.
66. Written Question P-1592/03 by Karin Scheele to the Commission, with answer *Subject: Legal admissibility of Member States' measures concerning the coexistence of crops*, <www.europarl.eu.int/bulletins/pdf/01c_bu-a(2003)06_en.pdf>
67. Katie Eastham and Jeremy Sweet, *Genetically Modified Organisms (GMOs): The significance of gene flow through pollen transfer* (European Environment Agency, 2002) pp.7–8.
68. Ibid.
69. Ibid., p.59.
70. Ibid. pp.59–60.
71. Karl Tolstrup et al. 'Report from the Working Group on the co-existence of genetically modified crops with conventional and organic crops: Conclusion and Summary' (Copenhagen: Ministry of Food, Agriculture

and Fish, January 2003), pp.6–7; James Randerson, 'Squeezed out: No room for GM crops in an "organic" Britain', *New Scientist*, 18 May 2002, p.14; 'Commission fails to lift the Moratorium', *Friends of the Earth Europe Biotech Mailout*, December 2003, p.1.

72. 'Limited EU role seen in GM crop management', *Environment Daily*, 5 March 2003.
73. 'EU Member States responsible for co-existence', *Friends of the Earth Biotech Mailout*, September 2003, p.1; 'The fight for GM free zones', *Friends of the Earth Biotech Mailout*, September 2003, p.10; European Commission press release, 'Commission rejects request to establish a temporary ban on the use of GMOs in Upper Austria', 2 September 2003; Tobias Buck, 'Brussels rejects Austria's modified crop ban', *Financial Times*, 3 September 2003.
74. 'No one will insure GM crops', FARM Press Release, 7 October 2003, <www.farm.org.uk/FM_Content.aspx?ID=138> FARM is a campaigning group set up to promote sustainable agriculture in Britain. Their survey found that insurers compared GMOs to thalidomide and asbestos, products which we were told were 'safe' when they were launched.
75. 'GMOs: The polluter does not pay', *Friends of the Earth Europe Bulletin*, March 2002.
76. 'Environment Council: Ministers start to water down liability directive', *European Report*, 8 March 2003.
77. Ibid.
78. 'EU secret study shows genetically engineered crops add high costs for all farmers and threaten organic', Greenpeace International Press Release, 16 May 2002; Anne-Katrin Bock et al., *Scenarios for Co-existence of Genetically-modified, Conventional and Organic Crops in European Agriculture* (Joint Research Centre of the European Commission, 2002) <www.jrc.cec.eu.int/GECrops/> 'EU's joint research centre says GMOs will mean financial losses for farmers', *Friends of the Earth Europe Biotech Mailout*, Vol. 8, Issue 3, June 2002, p.6.
79. European Parliament legislative resolution on the Council common position for adopting a European Parliament and Council directive on environmental liability with regard to the prevention and remedying of environmental damage (10933/5/2003 – C5–0445/2003 – 2002/0021(COD)), <www.europarl.eu.int/plenary/default_en.htm> At time of writing, the final text had yet to be agreed between Parliament and Council, but indications were that these amendments would be accepted; for a thorough examination of the Directive's inadequacies, see Marie-Martine Buckens (ed.) *Environmental Liability: Where the new EU Directive falls short* (Brussels: European Information Service, 2003).
80. European Federation of Biotechnology Task Group on Public Perceptions of Biotechnology *Briefing Paper 9: Biotechnology legislation in Central and Eastern Europe* (EFB, June 1999); Thomas Schweiger et al., *Will Enlargement Weaken EU Policy on GMOs?* (Friends of the Earth Europe/ Northern Alliance for Sustainability (FoEE/ANPED), 2003). This paper is the best guide to the state of play in the accession countries on the eve of enlargement. A copy can be obtained from the website at

<www.anped.org>; Iza Kruszewska *Romania: The dumping ground for genetically engineered crops: A threat to Romania's agriculture, biodiversity and EU accession* (Friends of the Earth Europe/Northern Alliance for sustainability /ECOSENS/Association of Organic Farmers Bioterra (FoEE/ ANPED/ECOSENS/Bioterra), 2003).

81. Vladimir Radyuhin, 'GM food battle moves to Russia', *The Hindu* (India), 17 September 2003.
82. Charles Kessler and Ioannis Economidis (eds) *EC Sponsored Research on Safety of Genetically Modified Organisms; A review of results*, <http://europa.eu.int/comm/research/fp5/eag-gmo.html>; 'Algae: biotechnology potential explored', *Europe Environment*, Section III, p.2, 26 February 2002; B. Zechendorf *Wonders of Life: Stories from Life Sciences Research (from the Fourth and Fifth Framework programmes)*, (European Commission Directorate General for Research/Office of Official Publications of the European Communities, 2002; S. Caro (ed.) *Ethical, Legal and Socio-economic Aspects of Agriculture, Fisheries and Food Biotechnology: An overview of research activities 1994–2002* (European Commission Directorate General for Research/Office for Official Publications of the European Communities, 2002).
83. European Commission *Life Sciences and Biotechnology: A strategic vision*, <http://europa.eu.int/comm/biotechnology/introduction_en.html>
84. Clive Cookson, 'The British experiment', *Financial Times*, 24 January 2001.
85. John Prideaux, 'Catholic nations may reject Brussels stem cell guide', *Financial Times*, 8 July 2003; Karen Carstens, 'Stem cell research is being "stonewalled" claims MEP', *European Voice*, 28 May 2003.
86. The European Group on Ethics in Science and New Technologies Commission of the European Communities, *General Report on the Activities of the European Group on Ethics in Science and New Technologies, 1998–2000* (Office for Official Publications of the European Communities, 2001), p.128; *Commission Staff Working Paper: Report on human embryonic stem cell research, SEC (2003 44)*, 3 April 2003; Eerste Kamer der Staten-Generall Vergaderjaar 2001–2002 Nr. 47c, 27, 423, *Wet houdende regels inzake handelingen met geslachtscellen en embryo's (Embryowet) (Lower House of the Dutch Parliament...Legally binding rules governing proceedings with sex cells and embryos (Embryo law))*; Clive Cookson and Victoria Griffith, 'Cloning alone: Britain looks set to overtake the US in biomedical research', *Financial Times*, 1 March 2002; Raphael Minder, 'Parliament approves EU stem cell research', *Financial Times*, 20 November 2003.
87. Before the Human Genome Project's discovery that human beings had far fewer genes than was expected, the paradigm in this area was known as 'Crick's Central Dogma'. This posited that each gene had one sole function, that it made a single protein. This is now known not to be the case, and current estimates have 30,000 genes somehow manufacturing 750,000 proteins. 'Genomics' is the study of the function of genes, but it is now clear that what is needed is a study of the function of each of these proteins: hence 'proteomics'.

88. Peter Mitchell, 'UK budget boosts biotech', *Nature Biotechnology*, Vol. 21, January 2003, p.9; Commission of the European Communities, *Commission Staff Working Paper: Report on human embryonic stem cell research*.
89. *EuropaBio's Biotechnology Information Kit. Module 16. Patenting* (the document is undated but appears to have been issued as part of the lobbying around the Biotechnology Directive, thus in 1997).
90. European Federation of Pharmaceutical Industries and Associations, Emerging Biopharmaceutical Enterprises, *Legal Protection of Biotechnological Inventions*, (EFPIA-EBE). The document is undated but from the contents it can de discerned that it was issued during the first half of 2001.
91. Daniel Alexander, untitled draft paper on patents on life forms, presented on 31 May 2001 to the European Parliament Temporary Committee on Human Genetics and Other New Technologies of Modern Medicine.
92. Nikolaus Thumm, 'Ethical considerations raised by biotechnology patents', *The IPTS Report*, No. 51, (Seville: IPTS, 2001), p.21.
93. Convention on the Grant of European Patents, Article 52(1); The Convention on the Grant of European Patents of 5 October 1973, its Implementing Regulations and Protocols as updated to 1995, are published as *European Patent Convention* (Munich: European Patent Office, 1995). The EPO has a website at <www.european-patent-office.org>
94. G. Van Overwalle, *The Legal Protection of Biotechnological Inventions in Europe and in the United States* (Leuven: Universitaire Pers, 1996) provides a detailed account of the system as it stood on the eve of the adoption of the Patents on Life Directive and following the updating of the EPC in 1995.
95. Proposal for a Council Regulation on the Community Patent, *Official Journal of the European Communities*, C337, 28 November 2000, <http://europa.eu.int/comm/internal_market/en/intprop/indprop/2k-714.htm>
96. Ibid.; Daniel J. Kevles *A History of Patenting Life in the United States with Comparative Attention to Europe and Canada* (European Group on Ethics in Science and New Technologies to the European Commission/Office for Official Publications of the European Communities, 2002); European Group on Ethics in Sciences and New Technologies to the European Commission, *Opinion on Ethical Aspects of Patenting Inventions Involving Human Stem Cells* (European Group on Ethics in Sciences and New Technologies to the European Commission/Office for Official Publications of the European Communities, 2002). For more on the European Group on Ethics in Science and New Technologies, an advisory body of the European Commission, see European Commission *The European Group on Ethics in Science and New Technologies*, November 2000, <http://europa.eu.int/comm/secretariat_general/sgc/ethics/en/index.htm>
97. Kevles, *A History of Patenting Life*, pp.59–60.
98. Ibid.

99. Geertrui van Overwalle, *Study on the Patenting of Inventions Related to Human Stem Cell Research*, (Luxembourg: European Group on Ethics in Science and New Technologies to the European Commission/Office for the Official Publications of the European Communities, 2002) p.41.
100. 'Proposal for a Council Directive on the Legal Protection of Biotechnological Inventions, 20 October 1988', *Official Journal of the European Communities*, C13, 13 January 1989.
101. Directive 98/44/ec of 6 July 1998 of the European Parliament and of the Council on the legal protection of biotechnological inventions, <http://europa.eu.int/eur-lex/en/lif/dat/1998/en_398L0044.html>
102. Ibid.; 'biological material' is defined earlier in the Directive (Article 2(1)(a)) as 'any material containing genetic information capable of reproducing itself or being reproduced in a biological system'.
103. Recital 21, Directive 98/44/EC.
104. Maggie Grace, 'A matter of life and death', *New Scientist*, 27 July 1996, p.60.
105. *Europe Environment*, 12 Mar 2002, Section III, p.5.
106. In 2000, following a 70 per cent increase over the previous five years in the number of firms active in the sector, it provided full-time employment for 8,000 scientists. Proportionate to population size, this put it – within the EU – behind only Sweden. Wil Thijssen, 'Het gevecht van onze genen', *De Volkskrant*, 15 July 2000.
107. Tweede Kamer der Staten general Vergaderjaar 2000–2001, 26 568 (R 1638) *Wijziging van de Rijksoctrooiwet…ten behoeve van de rechtsbescherming van biotechnologische uitvindingen* Nr 28 *Brief van de Staatsecretaris van Economische Zaken*, 21 November 2000 (letter from the Secretary of State for Economic Affairs to the Lower House of Parliament on the amendment of the patent law with regard to the protection of biotechnological inventions).
108. This would be very unusual, but it is possible under Article 173 of the Treaty on European Union.
109. Every legislative proposal emanating from the European Commission must be brought forward under a specified Article of the Treaty on European Union. If the Article were to be, in the view of the European Court of Justice, incorrectly applied, this would invalidate the Directive or other measure brought under it.
110. Van Overwalle, *Study on the Patenting of Inventions*, pp.48–9.
111. Sandy M. Thomas, Michael M. Hopkins and Max Brady, 'Shares in the human genome – the future of patenting DNA', *Nature Biotechnology*, Vol. 20, pp.1185–190, December 2002.
112. EFPIA-EBE, *Legal Protection of Biotechnological Inventions*.
113. 'A licence to print money', *New Scientist*, 16 May 1998, p.3; Andy Coghlan, 'The price of profit', *New Scientist*, 16 May 1998, p.20.
114. Fred Pearce. 'Greedy patenting could starve poor of biotech promise', *New Scientist*, 16 November 1996, p.6.
115. Francesco Fiori, European Parliament Temporary Committee on Human genetics and Other New Technologies of Modern medicine, Working Document on the social, legal, ethical, and economic implications

of human genetics, 8 June 2001, <www.europarl.eu.int/committees/genetics_home.htm>

116. Thomas et al., 'Shares in the human genome'.
117. Clive Cookson, 'Science and ethics clash over life forms', *Financial Times*, 2 September 2003.
118. Peter Mitchell, 'European Commission rethinks biotech patents', *Nature Biotechnology*, Vol. 20, pp.1175–76, December 2002.
119. See, for example, European Federation of Pharmaceutical Industries and Associations, Emerging Biopharmaceutical Enterprises, *From Industrial Property to Health-Related Enhancement: Directive 98/44/EEC* (EFPIA-EBE) Undated lecture notes probably issued in 2001.
120. Nell Boyce and Andy Coghlan, 'Your genes in their hands: Will patents on human genes encourage research or stifle it?', *New Scientist*, 20 May 2000, p.15.
121. Professor Piet Borst, ex-director of the Netherlands Cancer Institute, quoted in Marc van den Broek, 'Recht op genen', *De Volkskrant*, 6 November 2001, author's translation.
122. As things stand this is also controversial. A woman harbouring the defective gene would be by no means certain to develop breast cancer, and the only 'preventative measure' currently available is mastectomy, a somewhat drastic step to take purely as a precaution. This, however, is a separate issue.
123. Dolores Ibarreta and Nikolaus Thumm, 'Ethical Aspects of Biotechnological Patenting Revisited' *The IPTS Report*, 65 (Seville: Institute for Prospective Technological Studies, June 2002) p.19; European Parliament: Motions for resolutions on the patenting of BRCA1 and BRCA2 ('breast cancer') genes, B5–0633/2001, B5–0641/2001, B5–0651/2001, B5–0663/2001, 4 October 2001, These Resolutions are all available at <www.europarl.eu.int/comparl/tempcom/genetics/links/gene_links_ep.htm#1>; Laurence Frost, 'Breast cancer gene patent under fire from MEPs', *European Voice*, 2 May 2002.
124. Directive 98/44/EC of 6 July 1998.
125. In somatic cell nuclear transfer, the genetic material is removed from an egg and replaced with genetic material from the organism one is seeking to clone. As an embryo develops, it should have the same genetic features as the introduced material. This technique could be used to produce material genetically identical to the donor of the introduced genetic material and so could have great potential for producing replacement body parts (from microscopic components to whole organs) which would not be rejected by the patient's immune system. It could also, in theory, be used to clone entire people.
126. The European Parliament resolution on human cloning, adopted in 2000, asserts that 'there is no difference between cloning for therapeutic purposes and cloning for the purposes of reproduction' and calls on the member states 'to enact binding legislation prohibiting all research into any kind of human cloning', and, specifically, on the British Parliament to reject the measure then before it to permit research using embryos created by cell nuclear transfer. The Resolution, which reiterated and broadened earlier Resolutions (of 1993, 1997 and 1998) has no legal

force, but is merely an expression of the European Parliament's opinion. These Resolutions are all available at <www.europarl.eu.int/comparl/tempcom/genetics/links/gene_links_ep.htm#1>

127. Ibarreta and Thumm, 'Ethical Aspects of Biotechnological Patenting Revisited', p.21.
128. Patent EP 695351, Description, p.2, quoted in Van Overwalle, *Study on the Patenting of Inventions*, p.59.
129. Ibid.
130. European Parliament Resolution on the Decision by the European Patent Office with Regard to Patent No EP 695 351 Granted on 8 December 1999 ((Document BS-0288, issued on 30 March 2000), *Official Journal of the European Communities* L29, December 2000).
131. Communication from the Opposition Division of the EPO, point 5.2, quoted in van Overwalle, *Study on the Patenting of Inventions*, p.60; this document also contains a copy of the Edinburgh patent (pp.126–47). 'Somatic' refers to the ordinary, non-reproductive cells of the body. An 'enculeated oocyte' is an egg from which the genetic material has been removed.
132. Directive 98/44/EC of 6 July 1998, Article 8(2).
133. European Commission, *Report from the European Commission to the European Parliament and the Council on the Development and Implications of Patent Law in the Field of Biotechnology and Genetic Engineering,* 7/10/02, <http://europa.eu.int/eur-lex/en/com/rpt/2002/com2002_0545en01.pdf>
134. Quoted in Peter Mitchell, 'European Commission rethinks biotech patents'.
135. Nuffield Council on Bioethics, *The Ethics of Patenting DNA* (London: Nuffield Council on Bioethics, 2002).
136. Boyce and Coghlan, 'Your genes in their hands'; 'Genes off pat', *Financial Times*, 31 July 2002.
137. 'Bioethics report warns on number of DNA patents', *Financial Times*, 31 July 2002; 'Genes off pat', *Financial Times*, 31 July 2002; 'How many lawyers does it take...Greed and legal wrangling could stifle the medical revolution', *New Scientist*, 18 May 2002, p.3.
138. Commission on Intellectual Property Rights *Integrating Intellectual Property Rights with Development Policy* (London: DFID, 2002) cited in Thomas et al., 'Shares in the human genome', European Commission, *Report from the European Commission to the European Parliament and the Council on the Development and Implications of Patent Law in the Field of Biotechnology and Genetic Engineering.*
139. European Federation of Biotechnology Task Group on Public Perceptions of Biotechnology *Briefing Paper 1: Patenting in Biotechnology* (EFB, 1996), pp.2–3.
140. Ibid., p.4
141. 'Genetic Engineering: Transgenic patent compatible with EU rules', 29 January 2002, Section IV, p.5.

## CHAPTER 2


1. Organic Consumers' Association (US), *Consumer Warning: Genetically engineered hormones could be in your milk and dairy products*, <www.organicconsumers.org/text5.html>
2. Ibid.; Ronnie Cummins, *Monsanto's Genetically Engineered Products Meet Resistance*, <www.awionline.org/farm/rbgh.htm>
3. Organic Consumers' Association, *Consumer Warning*.
4. James Meek, 'Nature backs off GM crop claims', *Guardian*, 5 April 2002.
5. BBC Radio 4, *Seeds of Trouble*, first broadcast 7 January 2003, unofficial transcript prepared by Norfolk Genetic Information Network,<http://ngin.tripod.com/080103d.htm>; at the time of writing, the programme could be heard at <www.bbc.co.uk/science/seedsoftrouble.shtml>
6. For examples of such campaigners' reactions see the joint statement by Food First and others at <www.foodfirst.org/progs/global/ge/jointstatement2002.html> and the statement from Greenpeace, <www.greenpeaceusa.org/media/press_releases/2002/04242002.htm>; the letters themselves can all be read at <www.gene.ch/gentech/2002/Jun/msg00397.html>
7. See, for example, P. Christou, 'No credible scientific evidence is presented to support claims that transgenic DNA was introgressed into traditional maize landraces in Oaxaca, Mexico', *Transgenic Research*, Vol. 11, pp.iii–v.
8. Susan Milius 'Journal disowns transgene report', *Science News*, <www.phschool.com/science/science_news/articles/journal_disowns_transgene.html> (no date) 'Quist and Chapela reply', *Nature*, 416, 11 April 2002, p.602.
9. BBC Radio 4, *Seeds of Trouble*.
10. *Nature*, 11 April 2002; Alex Kirby, 'Doubts over Mexican GM maize report', BBC News (online), 5 April 2002; 'Astonishing denial of transgenic pollution', ISIS Report, 8 April 2002, <www.i-sis.org.uk>
11. J.E. Losey et al., 'Transgenic pollen harms monarch larvae', *Nature*, Vol. 399, 1999, p.214.
12. The findings of the six studies conducted to test Losey's findings were published in the *Proceedings of the National Academy of Sciences*, Vol. 98, pp.11908–42. See also a paper by two of the scientists involved in these studies, M.K. Sears and D. Stanley-Horn, 'Impact of Bt corn pollen on monarch butterfly populations', *Proceedings of the 6th International Symposium on the Biosafety of Genetically Modified Organisms*, <www.ag.usak.ca/isbr/Symposium/Proceedings/Section7.htm#120>
13. *Biotech Bias on the Editorial and Opinion Pages of major United States Newspapers and News Magazines*, (Food First/Institute for Food and Development Policy, 2002), <www.foodfirst.org/media/press/2002/biotechbiasreport.html>
14. Gregory Conko, 'Biotech critics find a little goes a long way', *Washington Times*, 23 June 2003.
15. A more detailed account of these events, as well as a great deal of other valuable information and links to websites expressing the full range

of views on this and related issues, is contained in the University of Colorado's Center for Life Sciences *Transgenic Crops: An introduction and resource guide*, at <www.colostate.edu/programs/lifesciences/TransgenicCrops/>

16. Vincent Kiernan, 'Truth is no longer its own reward', *New Scientist*, 1 March 1997.
17. Meryl Nass, 'Who is protecting the public health?', *Z Magazine*, April 2002.
18. Steven M. Druker, 'Advisory on US law and genetically engineered food: US food safety law mandates the precautionary principle' (Distributed to Members and staff of the European Parliament, July 2003, by the author and other activists from the Alliance for Bio-integrity).
19. Donna U. Vogt and Mickey Parish, *Food Biotechnology in the United States: Science, regulation, and issues*, Congressional Research Service (CRS), 1999, <www.usa.or.th/services/irc/gmo_crs.htm> pp.3–4.
20. Ibid., pp.2–4.
21. Recombinant DNA (rDNA) is DNA which has been formed by splicing together genes from different sources into new combinations.
22. Vogt and Parish, *Food Biotechnology in the United States*, p.7.
23. Jan-Peter Nap et al., 'The release of genetically modified organisms into the environment, Part 1: Overview of current status and regulations', *The Plant Journal*, Vol. 33, 2003, pp.1–18; Les Levidow, 'Bogged down in biotechnology's regulation', *New Scientist*, 13 June 1992, p.50.
24. *About the Food and Drug Administration*, <www.fda.gov/opacom/hpview.html>
25. Vogt and Parish, *Food Biotechnology in the United States*, pp.8–9.
26. *Washington Post*, 30 May 2003, quoted in Charles M. Benbrook, *GMOs, Pesticide Use, and Alternatives. Lessons from the US experience*, paper delivered at the conference on 'GMOs and Agriculture' held in Paris, 20 June 2003, p.15, posted at <www.biotech-info.net/lessons_learned.pdf>; Nap et al., 'The release of genetically modified organisms', p.9.
27. Vogt and Parish, *Food Biotechnology in the United States*, pp.8–9.
28. David Schubert, 'A different perspective on GM food', *Nature Biotechnology*, Vol. 20, No. 10, October 2002.
29. Vogt and Parish, *Food Biotechnology in the United States*, pp.8–9; Druker, 'Advisory on US law'. The FDA's own studies and discussions of GMOs can be found at <www.biotechnology.org>, the Alliance having, under the Freedom of Information Act, successfully sued the FDA for access to its files.
30. Vogt and Parish, *Food Biotechnology in the United States*, pp.9–10; Susan Katz Miller, 'Genetic first upsets food lobby', *New Scientist*, 28 May 1994, p.6.
31. Vogt and Parish, *Food Biotechnology in the United States*, pp.8–9; Martin Teitel and Kimberly A. Wilson *Changing the Nature of Nature: What you need to know about genetically engineered food* (London: Vision Paperbacks, 2000), p.68.
32. Kurt Kleiner, 'Soft words, big stick', *New Scientist*, 24 July 1999, p.12; 'GM industry turns on its latest critics', *Independent*, 15 July 1999;

'Meacher's GM charges rejected', BBC News (online), <http://news.bbc.co.uk/1/hi/uk_politics/3010254.stm>, 22 June 2003.

33. Kurt Kleiner, 'Farmers in the firing line', *New Scientist*, 25 September 1999, p.18.
34. Edward Alden, 'China to resume US soyabean imports', *Financial Times*, 8 March 2002; Lawrence Kojo Tsimese, 'Dumping GMOs in Africa', *Synthesis/Regeneration* 32, Fall 2003, <www.greens.org/s-r/32/32–05.html>
35. Letter from Food and Drug Administration Deputy Commissioner Lester Crawford to Governor John Kitzhaber of Oregon, quoted in Elizabeth Weise, 'FDA tries to remove genetic label before it sticks', *USA Today*, 9 October 2002.
36. Bill Lambrecht, 'Monsanto battles effort to require labeling of genetically modified food', *St Louis Post-Dispatch*, 19 September 2002; John A. MacDonald, 'Oregon drives debate on biotech-foods issue', *Portland Courant*, 29 August 2002.
37. Betty Martini quoted in Teitel and Wilson *Changing the Nature of Nature*, p.67; Andy Coghlan, 'Milk hormone data bottled up for years', *New Scientist*, 22 October 1994, p.4; Jim Ridgeway, 'Bush's ties to Monsanto: Franken Shrub', *The Village Voice*, January 2001, <www.villagevoice.com/issues/0104/ridgeway.shtml>; Robert Cohen, 'The Pelican Brief biotech: how big the fraud', <www.notmilk.com/pelican.html>; 'Some of the president's Monsanto men', *Idaho Observer*, February 2001, <http://proliberty.com/observer/20010208.htm>; ProdiGene Press Release, 'Anthony G. Laaos appointed to board for international food and agriculture development by President Bush', 9 September 2002.
38. The regulations can be read at <www.aphis.usda.gov/ppq/biotech/usregs.htm#usdalaw>; see also the comments in Consumers'Association, Policy Report, *GM Dilemmas – Consumers and genetically modified foods*, (London: Consumers' Association, 2002), p.91.
39. Nap et al., 'The release of genetically modified organisms', p.9; Ignacio Chapela, 'Farmers' Rights and the University-Industrial Partnership', lecture at the Berkeley City Council Chambers, Berkeley, California, 14 October 2002.
40. Vogt and Parish, *Food Biotechnology in the United States*, p.12.
41. Ibid., pp.11–12; Nap et al., 'The release of genetically modified organisms', p.9.
42. Nell Boyce, 'What's in a name…', *New Scientist*, 26 July 1997; Consumers' Association, Policy Report, *GM Dilemmas*, p.91; Andy Coghlan, 'US loosens laws for testing modified pest killers', *New Scientist*, 23 July 1994, p.10.
43. Federal Food, Drug and Cosmetic Act, as amended by the Food Quality Protection Act, 1996, Section 408.
44. 'Biotech's bitter harvest', *New Scientist*, 10 January 1998, p.3; Vogt and Parish, *Food Biotechnology in the United States*, pp.13–14.
45. TACD Press Release, 'EU and US consumer groups back EC initiatives on GMOs', 25 April 2002. For more information on TACD see <www.tacd.org>

46. Kleiner, 'Farmers in the firing line', p.18; 'Ignorance is bliss', *New Scientist*, 20 November 1999, p.25.
47. N.M. Chaudry and J.M. Regenstein, 'Implications of biotechnology and genetic engineering for kosher and halal foods', *Trends in Food Science Technology*, Vol. 5, 1994, pp.165–8; Vogt and Parish, *Food Biotechnology in the United States*, p.16.
48. 'Asda plans to go GM free – Asda has removed GM ingredients from own-label foods', BBC News (online), 26 January 2001, <http://news.bbc.co.uk/hi/english/uk/newsid_1137000/1137615.stm>
49. Charles M.Benbrook, *GMOs, Pesticide Use, and Alternatives*, pp.2–3; MacDonald, 'Oregon drives debate on biotech-foods issue'.
50. A full list is available at <www.fcpmc.com/english/events/presentations/>
51. Thomas Jefferson Hoban IV, *Biotechnology is Here to Stay: American retailers need not worry about consumer acceptance of foods produced with modern biotechnology*, and a video, *Biotechnology: Its role in your future*, <www4.ncsu.edu~hobantj/>; Karen Charman, 'The professor who can read your mind', *PR Watch* Vol. 6, No. 4, Fourth Quarter 1999, <www.prwatch.org/99-Q4/hoban.html#start>; 'Support for food biotechnology holds in the US', IFIC Press Release, 23 September 2002.
52. George Monbiot, 'Market enforcers: biotech firms found persuasion didn't work, so they are using a new tactic: coercion', *Guardian*, 21 August 2001.
53. Bill Freese *Manufacturing Drugs and Chemicals in Crops: Biopharming poses new threats to consumers, farmers, food companies and the environment*, <www.gefoodalert.org>
54. 'Food crops contaminated by bio-pharmaceuticals', *FoEE Biotech Mailout*, Vol. 8, Issue 6, December 2002, pp.7–8.
55. Justin Gillis, 'Soybeans mixed with altered corn: suspect crop stopped from getting into food', *Washington Post*, 13 November 2002; Associated Press report, 'FDA orders destruction of soybeans', 12 November 2002; Andrew Pollack, 'US investigating biotech contamination case', *New York Times*, 13 November 2002; Justin Gillis, 'Biotech firm mishandled corn in Iowa', *Washington Post*, 14 November 2002; Randy Fabi, 'US foodmakers urge ban on food crops for medicine', Reuters Securities News, 15 November 2002; Mike Toner, 'Tainted soybeans raise fears about "biopharming"', *Atlanta Journal-Constitution*, 17 November 2002; '"Iowa producers have a strong, scientific case for being involved in this new agricultural opportunity," says Sen. Grassley, who envisions biotechnology becoming a multimillion dollar industry in his state. "It's good to see that BIO has realized that they are putting unscientific restraints on Iowa and many other states."' Quoted in 'Puzzling industry response to ProdiGene fiasco', *Nature Biotechnology*, Vol. 21, January 2003, p.4; see also 'Puzzling industry response to ProdiGene fiasco', pp.3–4.
56. Friends of the Earth International Press Release, 'Mysterious "pharmaceutical" GM crop found in food chain', 15 November 2002.

57. Diane Carman, 'Bio-pharming: is cure worse than diseases?' *Denver Post*, 1 May 2003.
58. Ibid.; Peter Rosset, 'Anatomy of a "gene spill": do we really need genetically engineered food?', <www.foodforst.org/pubs/backgrdrs/2000/f00v6n4.html>
59. 'Here we go again', *New Scientist*, 6 July 2002, p.3.
60. Quoted in 'Puzzling industry response to ProdiGene fiasco', p.4.
61. Philip Cohen, 'Tougher laws on US pharm crops "not tough enough"', *New Scientist*, 22 March 2003, p.15.
62. Philip Cohen, Conference report, American Association for the Advancement of Science, 'Fighting over pharming', *New Scientist*, 1 March 2003, pp.22–3.
63. Ibid.
64. Ibid.
65. 'Slechte imago moet worden opgepoest' ('Bad image must be polished up'), *NRC Handelsblad*, 5 April 2000.
66. Daniel J. Kevles *A History of Patenting Life in the United States with Comparative Attention to Europe and Canada* (European Group on Ethics in Science and New Technologies to the European Commission/ Office for Official Publications of the European Communities, 2002), p.58.
67. Ibid., pp.4–13.
68. Ibid., p.14.
69. Ibid., pp.21–39; the case is referenced in Supreme Court records as *Diamond* v. *Chakrabarty*, 447 US 303, 100 S.CT, 2204 (1980), pp. 2211–12.
70. Kevles, *A History of Patenting Life in the United States*, p.45.
71. Ibid., pp.44–8.
72. 'Bigger and better', BBC website, <www.bbc.co.uk/science/genes/gene_safari/factfiles/beltsvillepig.html>
73. Quotes from Rep. Benjamin Cardin, Stewart Huber of the Wisconsin Farmers' Union and Tom Wagner, director of Edison Animal Biotechnology Center at University of Ohio, in Kevles *A History of Patenting Life in the United States*, pp.52–5.
74. Kevles, *A History of Patenting Life in the United States*, pp.56–7.
75. Ibid., pp.77–83.
76. Vincent Kiernan, 'Clinton smooths path for biotechnology', *New Scientist*, 7 January 1995, p.10; Andy Coghlan, '"Herd instinct" drove biotech firms to the brink', *New Scientist*, 21 October 1995, p.7.
77. Liz Else, 'The good fight', interview with Christopher Reeve, *New Scientist*, 15 March 2003, pp.54–6; 'Cloning votes', *Financial Times*, 15 March 2002; Clive Cookson, 'Most stem cell lines used in US research "will prove unviable"', *Financial Times*, 12 June 2002; Victoria Griffith, 'US plan for total cloning ban', *Financial Times*, 13 November 2002; Geoff Dyer, 'Norman Mailer lends weight to US anti-cloning coalition', *Financial Times*, 1 April 2002; Clive Cookson, 'US Charity plans $20m stem cell research', *Financial Times*, 18 November 2002.
78. Numerous accounts of Jesse Gelsinger's death have been written, and BBC's *Horizon* science documentary series produced a powerful and disturbing account of the tragedy, a summary description of which can

be read at <www.bbc.co.uk/science/horizon/2003/trialerror.shtml>; see also Sheryl Gay Stolberg, 'The biotech death of Jesse Gelsinger', *New York Times, Sunday Magazine*, 28 November 1999 and Nell Boyce, 'In memoriam: Tougher rules could be the legacy of gene therapy's first death', *New Scientist*, 18 December 1999.

79. Nell Boyce, 'Inquiry discovers "hidden" gene trial casualties', *New Scientist*, 12 February 2000, p.12.
80. See, for example, the account of the successful treatment of Rhys Evans in the *Great Ormond Street Hospital for Children Institute for Child Health Annual Report, 2001*, <www.ich.ucl.ac.uk/publications/annual_report_01_02/p36.html>
81. Nell Boyce, 'Stop the trials: activists demand a rethink on gene therapy', *New Scientist*, 18 March 2000.
82. Carl Elliott, 'Diary', *London Review of Books*, 28 November 2002.
83. Meryl Nass, 'Who is protecting the public health?', *Z Magazine*, April 2002.

## CHAPTER 3

1. *Monsanto* v. *Schmeiser*, <http://decisions.fct-cf.gc.ca/fct/2001/2001fct256.html>
2. 'Pollen furore', *New Scientist*, 18 October 2003, p.7.
3. The website of Alberta Agriculture (that is, the Canadian Province of Alberta's ministry of agriculture) at <www.agric.gov.ab.ca/crops/canola/outcrossing.html> gives a technical account of this, including estimates of the possible levels of contamination after a single season.
4. For more detail on the Schmeiser case, see E. Ann Clark, 'On the implications of the Schmeiser decision', *Bulletin of the Genetics Society of Canada*, June 2001, <www.plant.uoguelph.ca/faculty/eclark/percy.htm> and Niels Louwers and Marilyn Minderhoud, 'When a law is not enough: biotechnology patents in practice', *Biotechnology and Development Monitor*, No. 46, 2001, pp.16–19. Percy Schmeiser's battle continues, both in the courts and in the political arena: see Ed White, 'Schmeiser attempts Supreme Court', *The Western Producer*, 14 November 2002, <www.producer.com/articles/20021114/news/20021114news09.html>
5. Clive James, *Global Review of Commercial Transgenic Crops: 2000* (ISAAA Briefs, No. 23 2001), <www.isaa.org/publication/briefs/Brief_23.htm>
6. Consumers' Association, Policy Report, *GM Dilemmas – Consumers and genetically modified foods* (London: Consumers' Association, 2002), p.92; Canadian Food Inspection Agency website, <www.inspection.gc.ca/english/ppc/biotech/gen/statuse.shtml>
7. Devlin Kuyek, *The Real Board of Directors: The construction of biotechnology policy in Canada, 1980–2002,* (Sorrento, BC: The Ram's Horn, 2002) available online at <www.ramshorn.bc.ca> p.5.
8. Ibid., p.12.
9. Quote from an article published in the magazine *New Biotech*, May 1990, p.25, cited in Ibid. p.18.

10. Ibid., p.17. The other two were IT and 'advanced manufacturing materials'.
11. Ibid. pp.19–24.
12. James G. Heller Consulting, *Background Economic Study of the Canadian Biotechnology Industry* (paper commissioned by Industry Canada and Environment Canada, 1995), quoted in Ibid., p.14.
13. Ibid., pp.14–38, deals with this period.
14. Canadian Food and Drug Regulation, quoted in Consumers' Association, Policy Report, *GM Dilemmas*, pp.94–5.
15. Kuyek, *The Real Board of Directors*, pp.35–6.
16. Agriculture Canada, *Biotechnology Coordination in Agriculture Canada: Issues and activities for internal discussion* (Agriculture Canada internal document), October 1991, quoted in Ibid., p.36.
17. Ibid., p.38 Kuyek here cites a broader study of a phenomenon which apparently extended well beyond biotechnology and science policy and is fully explained in Donald J. Savoie, *Governing from the Centre: the concentration of power in Canadian politics* (University of Toronto Press, 1999).
18. Kuyek, *The Real Board of Directors*, pp.46–7.
19. Peter Kastner, 'Why we need to take care of R&D and why governments should help', a speech presented at a conference on industrial innovation at Simon Fraser University, Vancouver, 1999, available at <http://edie.cprost.sfu.ca/-grii/index.html>
20. Viola Sampson and Larry Lohmann, *Genetic Dialectic: The biological politics of genetically modified trees* (Sturminster Newton: The Corner House Briefing 21/Brighton: Econexus, 2000), <www.web-econexus.org> and <www.cornerhouse.icaap.org>, p.7.
21. Kuyek, *The Real Board of Directors*, p.41. Kuyek suggests (p.110) that Chrétien's enthusiasm for biotech might be partly due to the fact that his brother is a scientist specialising in protein chemistry, though, given the internationally evident identity of interests between neoliberal political-economic theory and biotech, such family connections, whilst to be expected, seem hardly necessary to an explanation of the importance placed on it by a government such as those which have prevailed in Canada since the Thatcher–Reagan era.
22. Ibid., p.41.
23. Ibid., pp.41–2. Genome Canada's website is at <www.genomecanada.ca>
24. Ibid.p.42.
25. Sampson and Lohmann, *Genetic Dialectic*, p.7.
26. Kuyek, *The Real Board of Directors*, pp.47–59, discusses the role of each ministry and the various federal ministers involved in more detail than would here be warranted, while pp.59–62 look at the government agencies responsible for grant awards, and pp. 62–6 look at the broader R&D picture. Pages 66–70 analyse the relationship between the rise in importance of 'Life Sciences' and the increasing privatisation of Canada's once-enviable health care system.
27. Ken Warn, 'Canada rules on muse patent', *Financial Times*, 6 December 2002; 'Biotech industry seeks legislation to permit patenting of

life', *Canadian Press* <www.globetechnology.com/servlet/story/RTGAM.20030415.gtbioapr15/GTStory>

28. Kuyek, *The Real Board of Directors*, p.74.
29. Ibid., p.75.
30. Compulsory licensing allows countries to authorise, under certain strict conditions, the production of an otherwise patent-protected drug if the company which produces it is unable to guarantee an affordable supply.
31. Stephen Leahy, 'Biotech hope and hype: the genetics revolution has failed to deliver', *Maclean's*, 30 September 2002.
32. General arguments for and against labelling reproduce more or less exactly the debate in the US and I have thus not quoted them here. For some Canadian examples see Gillian K. Hadfield, 'We need a label to identify genetically altered food', *Toronto Globe and Mail*, 10 May 1999; Barry Wilson, 'Canada afraid to upset US with GM labels', *Western Producer*, 21 November 2002; Consumers' Association, Policy Report, *GM Dilemmas*, pp.96–7.
33. Office of the Gene Technology Regulator, *Fact Sheet: The Gene Technology Regulator, the Ministerial Council and the Three Advisory Committees*; Health Australia, Gene Technology Ministerial Council, 10 September 2003, <www.health.gov.au/tga/gene/gtmc.htm>
34. Consumers' Association, Policy Report, *GM Dilemmas*, p.93.
35. Ibid., p.95. Details about the regulatory framework for gene technology in Australia (including information on the Act and Regulations) can be found at <www.health.gov.au/tga/genetech.htm> Office of the Gene Technology Regulator, *Fact Sheet: The GMO Regulatory System*; further information can also be found at the website of the Office of the Gene Technology Regulator, <www.ogtr.gov.au>
36. Office of the Gene Technology Regulator, *Fact Sheet: The record of GMO dealings and GM products* <www.ogtr.gov.au>
37. Ibid.
38. Office of the Gene Technology Regulator, *About the OGTR*, <www.ogtr.au/about/index.htm>
39. Office of the Gene Technology Regulator, *Information Bulletin: Federal Government provides a further two years funding for the Office of Gene Technology Regulator* (19 June 2003), <www.ogtr.gov.au/pubform/fundbulletin.htm>; Ian Lowe, 'Antipodes diary: planning for a genetic future', *New Scientist*, 8 July 2000; 'New Scientist Jobs: Australian biotech: Sun, surf and stem cells', *New Scientist*, 10 August 2002, pp.54–6.
40. Ian Lowe, 'Antipodes diary: making genes fit', *New Scientist*, 6 May 2000.
41. See, for example, 'A cautionary tale: Fish don't lay tomatoes', <www.aph.gov.au/senate/committee/clac_ctte/gene/report/contents.htm>
42. Senator Bob Brown, *Supplementary Report: The Gene Technology Bill 2000*.
43. Rachel Nowak, 'Australia agonises over stem cells', *New Scientist*, 21 September 2002, p.9.

44. Grant Holloway, 'Australia OKs embryo stem cell research', CNN Sydney, 4 December 2002, <www.cnn.com/2002/WORLD/asiapcf/auspac/12/04/australia.stemcells/>
45. The Growth and Innovation Strategy Biotechnology Framework can be read at <www.morst.govt.nz/uploadedfiles/Documents/Publications/Govt per cent20policy per cent20statements/BiotechFinal2.pdf>
46. </www.sustainabilitynz.org/docs/HSNOAct_ProposedAmendments.pdf> This document contains the text of the Act as originally adopted as well as amendments adopted before 2002 and those put forward after experience gained on implementation.
47. Background to the introduction of the labelling law can be found at the Ministry of Environment website, <www.mfe.govt.nz/issues/organisms/food.html>
48. This account of events leading up to the establishment of the Royal Commission on Genetic Modification relies particularly on two indigenous sources: C. Kay Weaver and Judy Motion, 'Sabotage and subterfuge: public relations, democracy and genetic engineering in New Zealand', *Media, Culture and Society*, Vol. 24, 2002, pp.325–43; and N. Legat, 'GM food: Why should we worry?', *North and South*, August 1999, pp.38–50, <www.gm.govt.nz/topics.shtml>
49. Lowe, 'Antipodes diary: making genes fit'; Kurt Kleiner, 'The accused; Is the biotech industry trying to silence one of its outspoken critics?', *New Scientist*, 3 March 2001, p.11.
50. Government of New Zealand, *Report of the Royal Commission on Genetic Modification* (2001), pp.2–3.
51. Ibid., pp.354–6.
52. Ibid.
53. Ibid., pp.356–60.
54. Government of New Zealand, *New Zealand Biotechnology Strategy: A foundation for development with care* (2003), p.1.
55. Ibid., pp.3–8.
56. Ibid., pp.10–13.
57. Ibid., p.16.
58. Ibid., pp.25–8.
59. Ibid., p.28.
60. European Commission, *Life Sciences and Biotechnology – A strategy for Europe: Communication from the Commission to the European Parliament, the Council, the Economic and Social Committee and the Committee of the Regions – COM (2002) 27* (Luxembourg: Office for Official Publications of the European Communities, 2002).
61. Government of New Zealand, *New Zealand Biotechnology Strategy*, p.29.
62. Terry Hall, 'Anger at NZ cow experiments', *Financial Times*, 2 October 2002.
63. Anne Beston, 'Report on GE maize crops savages agencies' role', *New Zealand Herald*, 20 December 2002; 'Maize report finds legal and accountability confusion', *Stuff*, 20 December 2002, <www.stuff.co.nz/stuff/0.2106.2145013a7693.00.html>

64. Government of New Zealand, *New Zealand Biotechnology Strategy*, p.29; Environment and Conservation Organisations of NZ, Media Release, 'Forest Research Institute GE tree planting unacceptable under international standards', 24 July 2003.
65. Government of New Zealand, *New Zealand Biotechnology Strategy*, p.34; note that p.35 of the *Strategy* contains a flow chart summarising biotechnology law in NZ.
66. The best website to follow the activities and views of opponents of genetic engineering in New Zealand is that of MADGE (Mothers Against Genetic Engineering) at <www.madge.net.nz/> To get the official story, the government view, and a summary of the legislation, go to <www.gm.govt.nz/topics.shtml>
67. Tracy Watkins, 'Privy Council axed, GM release approved', *Stuff*, 15 October 2003, <www.stuff.co.nz/stuff/0,2106,2692332a11,00.html>; David Fickling, 'New Zealand to allow trials of GM crops as two-year ban ends', *Guardian*, 30 October 2003.
68. European Federation of Biotechnology Task Group on Public Perceptions of Biotechnology, *Briefing Paper 8: Lessons from the Swiss biotechnology referendum* (EFB, 1998).
69. 'Swiss parliament edges towards GMO agreement', *Environment Daily*, 9 October 2002; 'GM-free food', *Swiss info*, <www.swissinfo.org/sen/Swissinfo.html?siteSect=113&sid=1396944>; Daniel Rechsteiner, 'Swiss GM moratorium initiative gains momentum', *Checkbiotech*, 15 January 2003.
70. 'Swiss moratorium back on the agenda', *Environment Daily*, 9 May 2003; 'Swiss MPs back-track over planned GM ban', *Environment Daily*, 13 June 2003; 'Genetic engineering: Swiss parliament rejects GM moratorium', *Europe Environment*, 27 June 2003, p.1.4; 'Clock ticking to Swiss anti-GM crop vote', *Environment Daily*, 22 September 2003.
71. United States Department of Agriculture, Foreign Agricultural Service, Global Agricultural Information Network (GAIN) Report, *Japan Biotechnology: Update on Japan's biotechnology safety approval and labeling policies*, (USDA, 2003); Consumers' Association, Policy Report, *GM Dilemmas – consumers and genetically modified foods* (Consumers' Association, 2002); the regulations can be read at <www.mhlw.go.jp/english/topis/food/>
72. 'Identity-preserved' systems have traditionally been used to guarantee, for example, that organic food is what it claims to be. It requires physical separation of a product at each stage of cultivation and shipping. Demand for IP non-GM food is now so high that it has spawned an entire industry. Private firms offer to guarantee supply of GM-free products, doing deals with farmers and other suppliers. There is even a handbook of non-GM sources, with an associated website, <www.non-gmosource.com/>
73. Consumers' Association, Policy Report, *GM dilemmas* (2002), pp.97–8; the regulations can be read at <www.mhlw.go.jp/english/topis/food/>
74. GAIN Report, *Japan Biotechnology*; an up-to-date list of MHLW approvals is maintained at <www.mhlw.go.jp/english/topis/food/sec01.html>

75. Aya Takada, 'Japan corn-trade paralysed by StarLink fear, higher costs', Reuters, 10 January 2003; 'Japan wheat buyers warn US on GE wheat', Organic Consumers' Association, 10 September 2003, <www.planetark.org/daily/newsstory.cfm/newsid/22173/story.htm>
76. David Pilling, 'Japan's changing climate starts to bear biotech fruit', *Financial Times*, 7 February 2001.
77. 'Single nucleotide variations single-nucleotide polymorphisms (SNPs) are single-base variations in the genetic code that occur approximately once every 1000 bases along the three billion bases of the human genome. Researchers believe that gaining knowledge of the locations of these closely-spaced DNA landmarks will ease both the sequencing of the genome and the discovery of genes related to such major human diseases such as asthma, diabetes, heart disease, schizophrenia and cancer.' Glossary of terms at <www.tmbioscience.com/glossary.php>
78. Pilling, 'Japan's changing climate starts to bear biotech fruit'.
79. Tade Matthias Spranger, 'The Japanese approach to the regulation of human cloning', *Zeitschrift für Japanisches Recht*, Heft Nr. 13/7. Jahrgang 2002, <www.djjv.org/japrecht/heft13/12-Spranger.htm>
80. Jiro Nudeshima, 'Human cloning legislation in Japan', <www.biol.tsukuba.ac.jp/~macer/EJ111/ej111b.htm>
81. Emma Young, 'Little dynamo heads off for biotech big time', *New Scientist*, 20 September 2003, pp.50–3; Alison Gee, 'Biotech firms take root in Singapore', *BBC World Business Report*, 8 August 2002; John Burton, 'Life sciences: Buying in the talents to create dynamic industries', Singapore supplement to *Financial Times*, 12 April 2002.
82. Young, 'Little dynamo heads off for biotech big time'; Wayne Arnold, 'Mice and more: Singapore's biotech drive', *New York Times*, 26 August 2003, <www.iht.com/articles/107756.html>
83. Young, 'Little dynamo heads off for biotech big time'.
84. Arnold, 'Mice and more'.
85. Ibid.
86. 'NBB Net: future plan', <www.nbbnet.gov.my/plan.htm>
87. Sang-Ki Rhee, *Biotechnology in Korea* (Korea Research Institute of Bioscience and Biotechnology, 2002).
88. Clive Cookson, 'S Korean scientists make breakthrough in cloning', *Financial Times*, 13 February 2004.
89. Joel I. Cohen, 'Harnessing biotechnology for the poor: challenges ahead for capacity, safety and public investment', *Journal of Human Development*, Vol. 2, No. 2, 2001, p.257, 'The case of China'.
90. Sylvia Pagán Westphal, 'First gene therapy approved', *New Scientist*, 29 November 2003.
91. Zhang-Liang Chen and Li-Jia Qu, 'The Status of agriculture biotechnology in China', <www.worldbiosafety.net/paper/01-Zhangliang per cent20Chen.doc>
92. Ibid.
93. 'China mulls first law to regulate biotech', Reuters report, 9 April 2002, <www.gene.ch/genet/2002/Apr/msg00054.html>
94. Peter Wonacott, 'China keeps biotech work quiet', *The Wall Street Journal Europe*, 21 January 2003.

95. Ibid.; Valerie Karplus, 'Let a thousand GM crops bloom' *International Herald Tribune*, 8 October 2003; Chen and Qu, 'The status of agriculture biotechnology in China'; 'China's caution on GM crops and foods', *AgBioIndia Mailing List* Editorial Comment, 30 October 2002; Joseph Kahn, 'The science and politics of super rice' *New York Times*, 22 October 2002.
96. Karplus, 'Let a thousand GM crops bloom'.

## CHAPTER 4

1. Devinder Sharma, 'The war of the world: America, GM, and developing countries', *OpenDemocracy*, 7 August 2003, <www.globalpolicy.org/socecon/ffd/2003/0807gmwar.htm>
2. Letter from Jeff Waage, International Organization of Biological Control, *New Scientist*, 29 March 1997, p.53.
3. United Nations Development Programme, *Human Development Report 2001: Making New Technologies Work for Human Development* (New York: UNDP, 2001), <www.undp.org/hdr2001>
4. Ibid.
5. Ibid.
6. European Commission, 'Giving rice an edge', <http://europa.eu.int/comm/research/conferences/2003/sadc/pdf/sac01–02.pdf>
7. Janice Wormworth, 'Enough to make everyone fat: Anuradha Mittal on GMOs and the Third World' (interview), *Link*, April 2000, p.19.
8. Nevin Scrimshaw, 'Golden Rice: blind ambition?' *Link*, April 2000, pp.12–13.
9. Fred Pearce, 'Cashing in on hunger: biotechnology's bid to feed the world is leaving less profitable techniques starved of funds', *New Scientist*, 10 October 1998.
10. European Commission, *Wonders of Life: Stories from Life Sciences Research from the Fourth and Fifth Programmes* (Office for Official Publications of the European Community, 2002), p.13.
11. 'Providing proteins to the poor – genetically engineered potatoes vs. Amaranth and pulses', Press Release from the Research Foundation for Science, Technology and Ecology, India, 9 January 2003; Andy Coghlan, '"Protato" to feed India's poor' *New Scientist*, 4 January 2003.
12. Friends of the Earth International, *Playing with Hunger: The reality behind the shipment of GMOs as food aid* (Friends of the Earth International, 2003) pp.4–5.
13. Ibid., pp.5–8; 'Would you feed this to your kids? Genetically modified food aid travels the globe', in *Clashes with Corporate Giants: 22 campaigns for biodiversity and community* (Amsterdam: Friends of the Earth International, 2002) pp.28–9.
14. Jonathan Matthews, 'The Fake Parade: under the banner of populist protest, multinational corporations manufacture the poor', *Environment*, 3 December 2002, <www.freezerbox.com/archive/article.asp?id=254>
15. James Lamont, 'Zambia turns away GM food aid for its starving', *Financial Times*, 19 August 2002.

16. Ibid.
17. 'UK Government Minister condemns "wicked" USAID GM food policy', Genetic Food Alert UK, <www.btinternet.com/~nlpwessex/Documents/sharm-tewolde.htm>
18. The Cartagena Protocol on Biosafety to the Convention on Biological Diversity, <www.biodiv.org/biosafety/default.aspx>
19. 'Force-feeding the hungry', Norfolk Genetic Information Network, October 2002, <http://ngin.tripod.com/forcefeed.htm>
20. 'BMA under attack', Norfolk Genetic Information Network, 31 January 2003; the article is partly an attempt to refute arguments put forward in Andy Coghlan, 'Zambia's GM food fear traced to UK', the text of which it also gives: <http://ngin.tripod.com/310103f.htm>
21. Letter from Andrew Clegg, Windhoek, Namibia, *New Scientist*, 31 August 2002, p.26.
22. Friends of the Earth International, *Playing with Hunger*, pp.10–11; Mae-Wan Ho, 'Africa unites against GM to opt for self-sufficiency', *Science in Society*, 16, Autumn 2002, pp.4, 9.
23. Friends of the Earth International, *Playing with Hunger*, pp.10–11;Vandana Shiva, 'Profits over people: how the World Food Summit in Rome last fortnight buried food rights, and clearly laid the contours of the future the powerful are designing', <www.flonnet.com/fl1913/19131180.htm>
24. James Lamont, 'Earth summit urged to focus on Africa food production', *Financial Times*, 19 August 2002.
25. 'US blames Europe's GMO ban for "letting people starve in Africa"', *European Voice*, 16 January 2003.
26. UK abc World Summit on Sustainable Development Daily Updates, 30 August 2002, 'Statement from leading Africans: why Africa should reject GE contaminated food' (UK abc is a group of development NGOs).
27. Emad Mekay, 'Plan will provide Africa with cheap biotech', Inter Press Service Report, 13 March 2003; 'Britain funds £13.4m GM programme in Third World', *Independent on Sunday*, 15 September 2002.
28. Maurizio Carbone, 'Biotech and the poor: a solution to the famine in Southern Africa?', *The Courier ACP-EU* No. 195, November–December 2002, pp.14–15.
29. Peter Newell, IDS Working Paper 201, *Biotech Firms, Biotech Politics: negotiating GMOs in India* (Brighton: Institute of Development Studies, 2003), Summary, p.iii.
30. 'India, US sign MoU to boost biotech sector', *Financial Express*, 8 November 2002, quoted in Newell, *Biotech Firms, Biotech Politics*, p.20.
31. Ibid.
32. Ibid., p.1.
33. Ibid., pp.3–4.
34. 'The Indian biosafety regulations on GMOs under test', <www.poptel.org.uk/panap/latest/test.htm>
35. Ibid.; Newell, *Biotech Firms, Biotech Politics*, p.3
36. Ibid., p.4

37. Ibid., p.6; Edward Luce, 'India approves growing of GM cotton', *Financial Times*, 27 March 2003.
38. Vandana Shiva, *Biotech Companies as Bioterrorists* (New Delhi: Research Foundation for Science, Technology and Ecology, undated document, c.2001); John Mason, 'Gaining ground', *Financial Times*, 27 March 2002.
39. *GE Crops in India*, <www.greenpeaceindia.org/gecrops.htm>
40. 'The Indian biosafety regulations on GMOs under test', <www.poptel.org.uk/panap/latest/test.htm>
41. Ibid.
42. See, for example, K.S. Jayaraman, 'India delays commercialization of GM mustard', *Nature Biotechnology*, Vol. 21, January 2003, p.9.
43. 'Syngenta pulls out of research collaboration with IGAU', Press Trust of India News, 10 December 2002.
44. Kurt Kleiner, 'Modified cotton passes 10-year trial with flying colours', *New Scientist*, 8 February 2003, p.21.
45. Habob Beary, 'Indian farmers target Monsanto', BBC news (online), <http://news.bbc. co.uk/1/hi/world/south_asia/3099938.stm> 11 September 2003; see also P.T. Bopanna, 'Protestors attack Monsanto greenhouse in southern India', Associated Press report, 9 November 2003.
46. Abhiram Ghadyal Patil Warora, 'Bollworm eats into Bt cotton's pride' *The Hitavada*, 10 October 2002; K. Venkateshwusu, 'Bt Cotton dashes hopes of ryots', *The Hindu*, 30 December 2002 (*ryot* = tenant farmer); 'Failure of Bt. Cotton in India', Press Release from the Research Foundation for Science, Technology and Ecology, New Delhi, 26 September 2002; 'Greenpeace demands probe into India gov'ts [sic] GM cotton claims', AFX Asia (England), 13 January 2003; Jaspal Singh Sidhu, 'Bt cotton failed in giving expected results say seed breeders', UNI report, 18 December 2002.
47. Suman Sahai, 'Pro-Agro's inferior GM mustard variety to be released soon', *Gene Campaign* (New Delhi), 23 September 2002; 'Tests on genetically modified mustard sought', *The Hindu*, 24 September 2002.
48. India Express Bureau 'India says 'no' to food aid from US', 2 January 2003, <http://indiaexpress.com/news/national/20030105-1.html>; Edward Luce, 'India rejects gene-modified food aid', *Financial Times*, 3 January 2003.
49. Joel I. Cohen, 'Harnessing biotechnology for the poor: challenges ahead for capacity, safety and public investment', *Journal of Human Development*, Vol. 2, No. 2, 2001, especially pp.249–53, 'Regulatory systems and the developing world: capacity and efficiencies'.
50. Ann Scholl and Facundo Arrizabalaga, 'Argentina: the catastrophe of GM soya', *Green Left Weekly* (Australia) 12 November 2003, p.20.
51. Stella Semino, 'The price Argentina has to pay for survival', unpublished report for the Grupo de Reflexión Rural Argentina-Europe, 20 October 2003.
52. Scholl and Arrizabalaga, 'Argentina: the catastrophe of GM soya', p.20.

53. Ibid.
54. Ibid.
55. Ibid.
56. Semino, 'The price Argentina has to pay for survival'.
57. Cohen, 'Harnessing biotechnology for the poor'.
58. Nadeem Iqbal, 'Pakistan opens doors to GM seed', *Asia Times,* 15 November 2002.
59. Presentation by Augusto Freire as part of 'The Cert-ID Non-GMO Traceability Conference', Antwerp, October 2003.
60. 'Brazil's U-turn', *New Scientist,* 4 October 2003.
61. John Mason, 'Gaining ground', *Financial Times,* 27 March 2002.
62. Raymond Colitt, 'Washington takes the battle over future for genetically modified crops to Brazil', *Financial Times,* 20 June 2003.
63. Caroline Daniel, 'Monsanto tales GM crusade to Brazil' *Financial Times,* 6 February 2003.
64. Elvino Bohn Gass, Member of the State Assembly, 'Rio Grande do Sul says no to transgenic crops' Press Release, 17 Feb 99; Reese Ewing, 'Lula government would favor GM-free Brazil', Reuters report, 3 October 2002; 'GE animal feed bad, GE-free feed good', Greenpeace website news reports, 13 September 2002, <www.greenpeace.org/news/details?news per cent5fid=26036>
65. Larry Rohter, 'In reversal, Brazil will permit a gene-modified crop', *New York Times,* 26 September 2003.
66. Ibid.
67. 'Brazil liberates GMOs', *Indymedia* report at <http://brasil.indymedia.org/en/blue/2003/09/264631.shtml>
68. ISNAR's website is at <www.cgiar.org/isnar/>
69. ISNAR, *Biotechnology at ISNAR* (The Hague, International Service for National Agricultural Research, 1999).
70. Ibid.
71. Morven A. Mclean et al., *ISNAR Briefing Paper 47: A conceptual framework for implementing biosafety: Linking policy, capacity and regulation* (The Hague, International Service for National Agricultural Research, March 2002).
72. Jose Falck-Zepeda et al., ISNAR Briefing Paper 54, *Biotechnology and Sustainable Livelihoods – Findings and recommendations of an international consultation* (The Hague, International Service for National Agricultural Research, 2002) p.2.
73. Ibid., p.7.
74. European Commission, *Creating Sustainable Solutions in Developing Countries* (Office for official publications of the European Communities, undated pamphlet).
75. Two collections of documents were available to those attending this conference, amongst whom this writer was unfortunately to be found. Entitled *Abstracts: Towards sustainable agriculture for developing countries: Options from life sciences and biotechnologies* and *Project Showcase: Towards sustainable agriculture for developing countries: Options from life sciences and biotechnologies,* they can be read at <http://europa.eu.int/comm/research/conferences/2003/sadc/index.html>

76. Food Ethics Council *Engineering Nutrition: GM crops for global justice?* (Food Ethics Council, 2003), pp.7–9.
77. Richard Hindmarsh, 'Constructing would-be molecular empires: behind the scenes', paper presented to the Peasant/Scientist Conference, Kuala Lumpur, 28–30 September 2003, available at <www.agbioindia.org/archive.asp>
78. Ibid.
79. Ibid.
80. Juan Lopez, 'The European-Latin American GMO debate: creating bridges to understanding', *Link*, April 2000, p.7.
81. Eusebius J. Mukhwana, 'Facing the food challenge: constraints and opportunities of using biotechnology in Africa', abstract distributed in Brussels at the European Commission's forum on biotechnology and development, February 2003.
82. Janice Wormworth, 'Biotech's endless race: interview with Miguel Altieri', *Link*, April 2000, p.9.

## CHAPTER 5

1. 'World Bank forges ahead with transgenic crops', Pesticide Action Network Updates Service Press Release, 27 September 2002, <http://sdnp.delhi.nic.in/resoyrces/biotech/news/et-11–2–00-icar.html>
2. United Nations Environment Programme, *UNEP International Technical Guidelines for Safety in Biotechnology* (Nairobi: UNEP, 1996), p.1.
3. Ibid., pp.1–4.
4. Aaron Cosbey and Stas Burgiel, *The Cartagena Protocol on Biosafety: An analysis of results* (International Institute for Sustainable Development (IISD), 2000), p.7.
5. The Cartagena Protocol on Biosafety to the Convention on Biological Diversity, Article 1. The text of the Protocol is available as well as background information is available at its official website, <www.biodiv.org/biosafety/default.aspx>
6. Jeffrey Waincymer, 'Cartagena Protocol on Biosafety' (unpublished legal analysis of the Protocol prepared for Deakin University, Melbourne, Australia, 2001), p.3, <www.waincymer2001.pdf>; The Cartagena Protocol on Biosafety to the Convention on Biological Diversity, Article 38.
7. Waincymer, 'Cartagena Protocol on Biosafety', p.3; The Cartagena Protocol on Biosafety to the Convention on Biological Diversity, Article 38; Susan George, 'Unfree trade: "GMOs are harmless, end of story"', *Le Monde diplomatique* (English-language edition), May 2002.
8. The Cartagena Protocol on Biosafety to the Convention on Biological Diversity, Article 1.
9. Ibid., Article 2.
10. Ibid.
11. Cosbey and Burgiel, *The Cartagena Protocol on Biosafety: An analysis of results*, p.9.
12. Ibid.

13. Ibid., p.10 and Preamble; Waincymer, 'Cartagena Protocol on Biosafety', p.4.
14. Cosbey and Burgiel, *The Cartagena Protocol on Biosafety: An analysis of results*, p.9.
15. Ibid. p.13.
16. Waincymer, 'Cartagena Protocol on Biosafety', p.14.
17. '...many U.S. government officials present at the negotiations worked with representatives from Argentina, Australia, Canada, Chile and Uruguay...' Donna U. Vogt and Mickey Parish, *Food Biotechnology in the United States: Science, regulation, and issues*, Congressional Research Service Report to Congress, 2 June 1999, p.22.
18. Peter Newell and Ruth Mackenzie 'The 2000 Cartagena Protocol on Biosafety: legal and political dimensions', *Global Environmental Change*, Vol. 10, 2002, pp.313–17.
19. Ibid.
20. Ibid.
21. Ibid.
22. Edward Alden, 'Greens and free-traders join to cheer GM crop deal', *Financial Times*, 31 January 2000.
23. A detailed report of the negotiations leading to the Protocol can be found at <www.iisd.ca/biodiv/excop/>
24. The Cartagena Protocol on Biosafety to the Convention on Biological Diversity, Article 7.
25. Ibid., Annex 1.
26. Ibid., Article 21.
27. Cosbey and Burgiel, *The Cartagena Protocol on Biosafety: An analysis of results*, p.7.
28. The Cartagena Protocol on Biosafety to the Convention on Biological Diversity, Article 7.
29. Ibid., Article 1.
30. Ibid., Article 12.
31. Cosbey and Burgiel, *The Cartagena Protocol on Biosafety: An analysis of results*, p.11.
32. Ibid., p.12.
33. The Cartagena Protocol on Biosafety to the Convention on Biological Diversity, Annex 2.
34. Ibid., Article 17.
35. Ibid., Article 18.
36. Ibid., Article 26.
37. Cosbey and Burgiel, *The Cartagena Protocol on Biosafety: An analysis of results*, p.8.
38. The Cartagena Protocol on Biosafety to the Convention on Biological Diversity, Article 11.
39. Ibid., Article 20.
40. Ibid., Article 22.
41. Ibid., Article 23.
42. Ibid., Article 18.2(a); Ashok B. Sharma, 'Cartagena Protocol, US Bioterror Act likely to influence trade', *Financial Express* (India), 3 November 2003; US Department of State, Office of the Spokesman, *Fact Sheet: The*

*Cartagena Protocol on Biosafety*, 16 February 2000, <http://usinfo.state.gov/topical/global/biotech/00021601.htm>

43. Basle Convention on the Control of Transboundary Movements of Hazardous Wastes and their Disposal, <www.unep.org>; J. Krueger, 'What's to become of trade in hazardous wastes?: the Basle Convention one decade later', *Environment*, November 1999, pp.10–21.
44. Cosbey and Burgiel, *The Cartagena Protocol on Biosafety: An analysis of results*, p.12.
45. Newell and Mackenzie, 'The 2000 Cartagena Protocol on Biosafety'.
46. 'Codex Task Force agrees on final draft of principles for the evaluation of GM foods', Food and Agriculture Organisation of the United Nations Press Release 02/24, 8 March 2002, <www.fao.org/WAICENT/OIS/PRESS_NE/english/2002/3060-en.html>; Frances Williams, 'UN body adopts global GM guidelines', *Financial Times*, 10 July 2003.
47. 'Modified food: Codex Alimentarius Agreement on Principles for Risk Analysis', *Europe Environment*, 26 March 2002, p.IV.6; Tam Dalyell, 'Westminster diary', *New Scientist*, 17 August 2002, p.58, quotes the then British Junior Health Minister, Hazel Blears, as describing the traceability issue as 'the main issue to be resolved by the task force'; Kimball Nill, 'A Codex Standard mandating "GMO Labeling" of food products containing genetically modified ingredients would decrease global food safety', paper distributed at the European Parliament, March 2001, by the American Soybean Association, of which Nill is Technical Director; Draft Principles for the Risk Analysis of Foods Derived from Modern Biotechnology, <www.codexalimentarius.net/biotech/en/ra_fbt.htm>; Draft Guideline for the Conduct of Food Safety Assessment of Foods Derived from Recombinant-DNA Plants, <www.codexalimentarius.net/biotech/en/DNAPlant.htm>
48. 'Experts agree global safety rules for GM foods', *Environment Daily* 1176, 12 March 2002; Williams, 'UN body adopts global GM guidelines', the full text of the agreement can be read at <www.codexalimentarius.net/ccfbt./bt02_01e.htm>
49. 'Boost for consumers as UN standards provide an agreed basis for GM food safety testing', letter from Anna Fielder, Director, Office of Developed and Transitional Economies, Consumers' International, *Financial Times*, 4 July 2003.
50. Crop Life International Press Release, 'Crop Life International launches new Reference Guide for biotech', 15 May 2003; Reference Guide for Biosafety Frameworks Addressing the Release of Plant Living Modified Organisms (Crop Life International, May 2003).
51. This draft text is available at <www.fao.org/ag/cgrfa/IU.htm>
52. Text of the Treaty and background materials are available at <www.fao.org/ag/cgrfa/itpgr.htm>; comment on the Treaty can be found through the Genomics Gateway at <www.bradford.ac.uk/acad/sbtwc/gateway/TRADE/ITPGR.htm>; see also *Summary Report of the Roundtable on Plant Genetic Resources in Africa's Renewal, held in Nairobi, Kenya from 2 to 3 April 2000*, <http://web.idrc.ca/uploads/user-S/10545704940ROUNDTREP-edited_FIN.doc>

53. Clive Cookson, 'Funding accord over stem cells', *Financial Times*, 9 January 2003.
54. Mark Turner and Clive Cookson, 'Move to ban human cloning likely to divide UN members', *Financial Times*, 29 October 2003; 'UN derails ban on human cloning', BBC News (online), 7 November 2003, <http: newsvote.bbc.co.uk/mpappa/pagetools/print/news.bbc.co.uk/2/hi/science/nature/>; Mark Turner, 'Close UN vote postpones decision on banning human cloning for two years', *Financial Times*, 7 November 2003.
55. 'Bush presses for human cloning ban', BBC News (online), 10 April 2002, <http://news.bbc.co.uk/hi/english/sci/tech/newsid_1922000/1922211.stm>
56. Turner, 'Close UN vote postpones decision on banning human cloning for two years'.
57. Universal Declaration on the Human Genome and Human Rights, text available at <www1.umn.edu/humanrts/instree/Udhrhg.htm>
58. Ibid., Article 2.
59. Ibid., Article 4.
60. Ibid., Article 5a, 5b, 5c.
61. Ibid., Article 5e.
62. Ibid., Article 6.
63. Ibid., Articles 7, 8, 9.
64. Ibid., Article 11.
65. Ibid., Article 12.
66. Ibid., Article 17.
67. Ibid., Article 18.
68. See the Biological and Toxic Weapons Convention website, <www.opbw.org/>
69. Frances Williams and Stephen Fidler, 'Bioweapons talks reopen amid discord on verification', *Financial Times*, 11 November 2002.
70. Ibid.
71. Frances Williams, 'Bioweapon move backed', *Financial Times*, 15 November 2002.

## CONCLUSION

1. See, for example, Stanley W.B. Ewen and Arpad Pusztai, 'Effect of diets containing genetically modified potatoes expressing *Galanthus nivalis* lectin on rat small intestine', *The Lancet*, Vol. 354, Issue 9187, 1998, p.1353.
2. 'How safe is GM food?', *The Lancet*, Vol. 360, No. 9342; since this was written, the Codex Alimentarius Commission has gone some way to meeting this demand.
3. Stephen Nottingham, *Genescapes: The ecology of genetic engineering* (London; Zed Books, 2002), p.12.
4. Quoted in *New Scientist*, 31 August 2002, p.41.
5. The industry body EuropaBio estimates the growth (in hectares globally) as follows: 1996, 1.7 million; 1997, 11 million; 1998, 27.8 million;

1999, 39.9 million; 2000, 44.2 million, EuropaBio *GMO Fact Sheet*, 2001, source acknowledged as Clive James, *Global Status of Commercialized Transgenic Crops 2000, ISAA Brief No. 21* (2000).

6. G.J. Persley *New Genetics, Food and Agriculture: Scientific discoveries – societal dilemmas* (International Council for Science (ICSU), 2003), pp.16–17, <www.icsu.org> This is true up to the end of 2003. More products may of course be licensed after that.
7. ISAAA's goal is to promote agricultural biotechnology in developing countries. For more information, see International Service for the Acquisition of Agri-biotech Applications website at <www.isaaa.org/>
8. ETC Group Communique, *Globalization, Inc*, Issue 71, July/August 2001, quoted in Devlin Kuyek, *The Real Board of Directors: The construction of biotechnology policy in Canada, 1980–2002* (Sorrento, BC: The Ram's Horn, 2002), p.99. Kuyek notes that he has adjusted ETC's figures, published a year before his book, to take into account 'Bayer's purchase of the agricultural division of Aventis and Aventis' controlling interests in the seed companies Groupe Limagrain and KWS AG'.
9. Daniel Charles, *Lords of the Harvest: Biotech, big money, and the future of food* (Cambridge, MA: Perseus Publishing, 2002) p.180.
10. United States Department of Agriculture Economic Research Service, *Adoption of Bioengineered Crops,* (2001), p.21.
11. Ma-Wan Ho and Joe Cummins, 'Failures of gene therapy', *Science in Society,* Issue 16, Autumn 2002, p.13; Carol Ezzell, 'Move over, human genome', *Scientific American*, April 2002, p.28.
12. *Information Paper on Ethical, Social and Public Awareness Issues in Gene Therapy* (EuropaBio, 2000) p.3; *Beleidsnota Biotechnologie, Vergaderjaar 2000–2001*, Tweede Kamer der Staten-Generaal, Nederland (author's translation); Andrew Webster et al., *Human Genetics: An inventory of new and potential developments in human genetics and their possible uses* (European Parliament Directorate General for Research, Science and Technological Options Assessment (STOA), 2001), p.34. See also the accounts of two practical applications of gene therapy, one of which took a life while the other saved one, at <www.bbc.co.uk/science/horizon/2003/trialerror.shtml>; Sheryl Gay Stolberg, 'The biotech death of Jesse Gelsinger', *New York Times, Sunday Magazine*, 28 November 1999, and Nell Boyce, 'In memoriam: tougher rules could be the legacy of gene therapy's first death', *New Scientist,* 18 December 1999; *Great Ormond Street Hospital for Children Institute for Child Health Annual Report, 2001*, <www.ich.ucl.ac.uk/publications/annual_report_01_02/p36.html>. Other useful accounts of and comments on gene therapy can be found at Brian J. Ford, *Genes: The fight for life* (London: Cassell, 1999) p.2; for examples of apparently promising lines of gene therapy research, see Sylvia Pagán Westphal, 'Cure for deafness comes a step closer', *New Scientist,* 7 June 2003; Victoria Griffith, 'Haemophiliacs see new hope in gene therapy', *Financial Times,* 16 February 2002; 'Haemophilia treatment hope', BBC News (online), <http://news.bbc.co.uk/hi/english/health/newsid_2030000/2030051.stm>; '"Scrunched" DNA aids gene patients', BBC News (online), 8 May 2002, <http://news.bbc.co.uk/hi/english/health/newsid_1976000/1976113.stm>;

'"Spy virus" could aid cancer fight', BBC News (online), 9 May 2002, <http://news.bbc.co.uk/hi/english/health/newsid_1974000/1974974.stm>

13. Clive Cookson, 'Nobel scientist goes back to the lab' (interview with Sir Paul Nurse), *Financial Times*, 6 March 2003.
14. Ian Wilmut, Keith Campbell and Colin Tudge, *The Second Creation: The age of biological control by the scientists who cloned Dolly* (London, Headline, 2002) p.312.
15. Marcy Darnovsky, 'Genetics and society: the misstep of human cloning', *San Francisco Chronicle*, 6 January 2003, <www.sfgate.com/cgi-bin/article.cgi?file=/chronicle/archive/2003/01/06/ED3915.DTL>
16. Roger Dobson, 'Clinic to offer couples test for embryo defects', *Sunday Times*, 1 September 2002.
17. Prof. Jean-Louis Mandel, Institut de Génétique et de Biologique moléculaire et cellulaire, Strasbourg, speaking at a dinner debate in Strasbourg, 11 December 2001, published with other contributions as *New Technologies and Health Care* (Paris: Institut des Sciences de la Santé, 2002) p.6; Sylvia Pagán Westphal, 'Your very own sequence', *New Scientist*, 12 October 2002, pp.12–13.
18. 'Gene testing is easy, it's the next bit that's hard', *New Scientist*, 3 May 2003, p.3.
19. Bob Holmes, 'The code breakers', *New Scientist*, 9 November 2002, pp.56–7.
20. Karen Schmidt, 'Just for you: one person's cure can be somebody else's poison', *New Scientist*, 14 November 1998, p.32; Webster et al., *Human Genetics*, p.23.
21. Webster et al., *Human Genetics*, p.24.
22. Ibid.

# Index